FIFTH EDITION

EXERCISES IN HELPING SKILLS

A Manual to Accompany *The Skilled Helper*

FIFTH EDITION

EXERCISES IN HELPING SKILLS
A Manual to Accompany *The Skilled Helper*

Gerard Egan
Loyola University of Chicago

Brooks/Cole Publishing Company
Pacific Grove, California

ITP™ The trademark ITP is used under license.

A CLAIREMONT BOOK

Brooks/Cole Publishing Company
A Division of Wadsworth, Inc.

Printed in the United States of America

10   9   8   7   6   5   4   3

ISBN 0-534-21295-6

Sponsoring Editor: **Claire Verduin**
Marketing Representative: **Thomas L. Braden**
Editorial Associate: **Gay C. Bond**
Production Coordinator: **Dorothy Bell**
Cover Design and Illustration: **Katherine Minerva**
Printing and Binding: **Malloy Lithographing, Inc.**

CONTENTS

PART ONE

LAYING THE GROUNDWORK

The exercises in this manual are meant to accompany the fifth edition of *THE SKILLED HELPER* by Gerard Egan (Brooks/Cole Publishing Company, Pacific Grove, California, 1994). The four parts of the manual — I. Laying the Groundwork, II. Basic Communication Skills for Helpers, III. Stage I of the Helping Model and Advanced Communication Skills, and IV. Helping Clients Develop Programs for Constructive Change — correspond to the four parts of the text. The "sections" of the manual correspond to the chapters of the book. The exercises themselves are numbered consecutively throughout the text.

Section 1
INTRODUCTION

You may or may not intend to become a professional helper. Whether you do or not, learning the model, methods, and skills of *The Skilled Helper* can help you become more effective in interactions between yourself and others in all the social settings of life including family, friendship groups, and work settings.

   The sections dealing with what helping is all about and the goals of helping are the most important parts of Chapter One. It is important that you understand what helping is all about in order to get the most out of doing these exercises.

EXERCISE 1: UNDERSTANDING WHAT HELPING IS ALL ABOUT

1. Read the sections dealing with what helping is all about, the goals of helping, and helping as a learning process.
2. Picture yourself talking with someone with whom you are about to establish a helping relationship.

3. In your own words describe to that person what you see helping to be.

_____

_____

_____

_____

_____

_____

_____

_____

_____

_____

4. Share what you have written with a learning partner. Note the similarities and differences between the two statements. To what degree do they include the main ideas outlined in the text? In what way do you think you should modify your statement?

THE PURPOSE OF THESE EXERCISES

Against the background of what helping is all about, the exercises in this manual serve a number of purposes:

1. **A Behavioral Grasp.** They can help you develop a **behavioral** rather than merely a cognitive grasp of the principles, skills, and methods that turn helping models into useful tools. They give you a "feel" for methods and skills before you use them in interactions with others.

2. **Self-Exploration.** They can be used to help you explore your own strengths and weaknesses as a helper. That is, they provide a way of having you apply the helping model to **yourself** first before trying it out on others. As such, they can help you confirm strengths that enable you to be with clients effectively and manage weaknesses that would stand in the way of helping clients manage problem situations.

3. **Personal Problem Solving.** You can use these exercises to become a better problem manager and opportunity developer in your own life.

4. **Client Participation.** You can help clients use these exercises selectively to explore and manage their own problems in living more effectively. These exercises provide one way of promoting client participation in the helping process.

5. **Training Clients in Problem Management.** You can help clients use these exercises to learn the **skills of problem management** themselves. Training in problem-management skills encourages self-responsibility in clients and helps make them less dependent on others in managing their lives.

A TRAINING PROGRAM FOR HELPERS

The following are standard steps in a skills-training program:

1. **Cognitive Understanding.** Develop a cognitive understanding of a particular helping method or the skill of delivering it. You can do this by reading the text and listening to lectures.

2. **Clarification.** Clarify what you have read or heard. This can be done through instructor-led questioning and discussion.

The desired outcome of Steps 1 and 2 is **cognitive clarity.**

3. **Modeling.** Watch experienced instructors model the skill or method in question. This can be done "live" or through films and videotapes.

4. **Written Exercises.** Do the exercises in this manual that are related to the skill or method you are learning. The purpose of this initial use of the method or skill is to demonstrate to yourself that you understand the helping method or skill enough to begin to practice it. The exercises in this manual are a way of practicing the skills and methods "in private" before practicing them with your fellow trainees. They provide a behavioral link between the introduction to a skill or method that takes place in the first four steps of this training format and actual practice in a group.

The desired outcome of Steps 3 and 4 is **behavioral clarity.**

5. **Practice.** Move into smaller groups to practice the skill or method in question with your fellow trainees.

6. **Feedback.** During these practice sessions, evaluate your own performance and get feedback from a trainer and from your fellow trainees. This feedback serves to confirm what you are doing right and to correct what you are doing wrong. The use of video to provide feedback is also helpful.

The desired outcome of Steps 5 and 6 is **initial competence**
in using the model and in the skills that make it work.

7. **Evaluating the Learning Experience.** From time to time stop and reflect on the training process itself. Take the opportunity to express how you feel about the learning program and how you feel about your own progress. While Steps 1 through 6 deal with the task of learning the helping model and the methods and skills that make it work, Step 7 deals with group maintenance, that is, managing the needs of individual trainees. Doing this kind of group maintenance work helps establish a learning community.

8. **Supervised Practice with Actual Clients.** Finally, when it is deemed that you are ready, apply what you have learned to actual clients. Supervision is an extremely important part of the learning process. Indeed, effective helpers never stop learning about themselves, their clients,

and the helping process itself.

The desired outcome of Steps 7 and 8 is **proficiency**.

The program in which you are enrolled may cover only a few of these steps. Comprehensive programs for training professional helpers must eventually include all steps.

DIFFERENT WAYS OF USING THESE EXERCISES

The way you use these exercises depends on the kind of training program you are in. The approach may be academic or experiential.

- **Academic Programs**. If the program is more or less academic in nature, then you can use these exercises to provide yourself with a *behavioral* feel for the model, methods, and skills involved in skilled helping. In that case, you would skip the exercises or the parts of the exercises that demand interaction with others. Even if your program has an academic cast to it, you might want to get a learning partner and share the learnings you glean from these exercises. One way of learning the stages and steps of this helping model is to apply them to your own problems and concerns first. Applying what you are learning to the problems and unused opportunities in your own life is often a distinct benefit.
- **Experiential Training Programs**. If your program is experiential in nature, then these exercises can help you prepare for involvement with your co-learners. In experiential programs the instructor usually coordinates reading, doing these exercises, and practice involving other members of the group.

There are various approaches in an experiential program. You may role play certain kinds of clients or you may work on your own real issues or you may do a combination of both.

- **Role Playing**. In experiential programs you are going to be asked to act both as helper and as client in practice sessions. In the written exercises in this manual, you are asked at one time or another to play each of these roles. Role playing is not easy. It forces you to get "inside" the clients you are playing and understand their concerns as they experience them. This can help you become more empathic.
- **Dealing with Real Issues**. Role playing, while not easy, is still less personally demanding than discussing your own real concerns in practice sessions. Under certain conditions, the training process can be used to look at some of the real problems or concerns in your own life, *especially issues that relate to your effectiveness as a helper*. For instance, if you tend to be an impatient person — one who places unreasonable demands on others — you will have to examine and change this behavior if you want to become an effective helper. Or, if you are very nonassertive, this may assist you in helping clients challenge themselves.

Another reason for using real problems or concerns when you take the role of the client is that it gives you some experience of **being** a client. Then, when you face real clients, you can appreciate some of the misgivings they might have in talking to a relative stranger about the intimate details of their lives. Other things being equal, I would personally prefer going to a helper who has had some experience in being a client.

If what you find out about yourself through these exercises is not to be shared with others, then dealing with real issues is not a concern. However, if you are going to talk about personal issues in the training sessions, you should do so only under certain conditions:

- **A Safe and Productive Training Group.** Dealing with personal concerns in the training sessions will be both safe and productive if you have a competent trainer who provides adequate supervision, if the training group becomes a learning community that provides both support and reasonable challenge for its members, and if you are willing to discuss personal concerns. Self-disclosure will be counterproductive if you let others extort it from you or if you attempt to extort it from others. Your self-disclosure should always remain appropriate to the goals of the training group. Extortion, "secret-dropping," and dramatic self-disclosure are counterproductive.
- **Adequate Preparation for Self-Disclosure.** If you are to talk about yourself during the practice sessions, you should take some care in choosing what you are going to reveal about yourself. Making some preparation for what you are going to say can prevent you from revealing things about yourself that you would rather not discuss in a training group.

The self-exploration exercises in Appendix One will help you discover and explore the kinds of concerns that you can safely deal with in an experiential training group.

A CAUTION: EXERCISE MANUAL AS TOOL

This manual should be used to the degree that it helps you develop a working knowledge and behavioral feel for essential helping methods and skills. It is not an end in itself. Trainees learn in different ways and at different rates. If you already have a particular skill, then working through an exercise to acquire that skill will be tedious and self-defeating. If doing two items in an exercise is enough to give you an understanding and behavioral feel for the method or skill in question, then doing ten items will prove to be equally tedious and self-defeating. In sum, doing these exercises just to do them could leave a bad taste in your mouth for helping itself.

Throughout this manual you are urged to share your responses and get feedback from a learning partner. In classes based on an experiential learning approach, the instructor often provides a "learning from one another" structure. But if you are using these exercises on your own in a more academic approach or if an interactive structure is not provided, then your learning through these exercises will be enhanced if you find a compatible learning partner or join a "self-managed" group of learners. Learning helping skills is as much a social exercise as is helping itself. Finding a forum in which you can explore these exercises in a give-and-take fashion is an essential enhancement to your learning.

GIVING FEEDBACK TO SELF AND OTHERS

If you are working with a learning partner or if you are in an experiential training group, you will be called on to give feedback both to yourself and to your co-learners on how well you are learning and using helping methods and skills. Giving feedback well is an art. Here are some guidelines to help you develop that art.

1. **Keep the goal of feedback in mind.** In giving feedback, always keep the generic goal of feedback in mind: To help the other person (or yourself, in the case of self-feedback) do a better job. Improved performance is the goal. Applied to helping, this means providing the kind of feedback to yourself and your co-learners that will help you become better helpers. Feedback will help you learn every stage and step of the model.

2. **Give positive feedback.** Tell your co-learners what they are doing well. This reinforces useful helping behaviors. "You leaned toward him and kept good eye contact even though he became very intense and emotional. Your behavior sent the right 'I'm still with you' message."

3. **Don't avoid corrective feedback.** To learn from our mistakes we must know what they are. Corrective feedback given in a humane way is a powerful tool for learning. "You seem reluctant to challenge your clients. For instance, Sam [the client] didn't fulfill his contract from the last meeting, and you let it go. You fidgeted when he said he didn't get around to it."

4. **Be specific.** General statements like "I liked your style in challenging your client" or "You could have been more understanding" are not helpful. Change them to statements such as, "Your challenge was helpful because you pointed out how self-defeating her internal conversations with herself are and hinted at ways she could change these conversations." Or, "Your tone was harsh, and you did not give him a chance to reply to what you were saying."

5. **Focus on behavior rather than traits.** Point out what the helper does or fails to do. Do not focus on traits or use labels such as, "You showed yourself to be a leader" or "You're still a bumbler." Avoid using negative traits such as "lazy," "a slow learner," "incompetent," "manipulative," and so forth. This is just name calling and creates a negative learning climate in the group. The following statements deal with specific behaviors rather than traits: "You let him criticize you without becoming defensive; he listened to you better after that" or "You did not catch her core message; in fact, you seem to have difficulty in listening well enough to catch your clients' core messages." Such statements deal with specific behaviors rather than traits.

6. **Indicate the impact of the behavior on the client.** Feedback should help counselors-to-be interact more productively with clients. It helps, then, to indicate the impact of the helper's behavior on the client. "You interrupted the client three times in the space of about two minutes. After the third time, she spoke less intensely and switched to safer topics. She seemed to wander around."

7. **Provide hints for better performance.** Often, once helpers receive corrective feedback, they know how to change their behavior. After all, they are learning how to help in the training program. Sometimes, however, if the helper agrees with the feedback but does not know how to change his or her behavior, suggestions or hints on how to improve performance are useful. These, too, should be specific and clear. "You are having trouble providing your clients with empathy because you allow them to talk too long. When they go on and on, they make so many points that you don't know which to respond to. Try 'interrupting' your clients gently so you can respond to key messages as they come up."

8. **Be brief.** Feedback that is both specific and brief is most helpful. Long-winded feedback proves to be a waste of time. A helper might need feedback on a number of points. In this case, provide feedback on one or two points. Give further feedback later on. In general, do not overload your co-learners with too much feedback at one time.

9. **Use dialogue.** Feedback is more effective if it takes place through a dialogue between the giver and receiver, a brief dialogue, of course. This gives the receiver an opportunity to clarify what the feedback giver means and to ask for suggestions if he or she needs them. A dialogue helps the receiver better "own" the feedback.

Section 2
THE STAGES AND STEPS OF THE HELPING PROCESS

Here is a brief outline of the skilled-helper model. It focuses on problem management and opportunity development. In order to see this model illustrated through a case, read Chapter Two of *The Skilled Helper* (5th edition, 1994).

STAGE I: THE CURRENT SCENARIO
"WHAT'S GOING ON?"
Helping Clients Explore Problems and Unused Opportunities

Clients can neither manage problem situations nor develop unused opportunities unless they identify and understand them. Exploration and clarification of problems and opportunities take place in Stage I. This stage deals with the current state of affairs, that is, the problem situations or unused opportunities that prompt clients to come for help. This stage includes the following steps:

1. **Help clients tell their stories.** First of all, clients need to talk about their problems and concerns, that is, tell their stories. Some do so easily, others with a great deal of difficulty. You need to develop a set of attitudes and communication skills that will enable you to help clients reveal problems in living and unused potential. This means helping clients find out what's going wrong and what's going right in their lives. Successful assessment helps clients identify both problems and resources. It also helps clients spell out problem situations in terms of specific experiences, behaviors, and feelings.

2. **Help clients challenge their blind spots and develop new perspectives.** This means helping clients see themselves, their concerns, and the contexts of their concerns more objectively. This enables clients to see more clearly not only their problems and unused opportunities, but also ways in which they want their lives to be different. Your ability to help clients challenge themselves throughout the helping process adds a great deal of value.

3. **Help clients focus on substantive problems.** This means helping clients identify their most important concerns, especially if they have a number of problems. Effective counselors help clients work on "high-leverage" issues, that is, issues that will make a difference in clients' lives.

STAGE II: DEVELOPING A PREFERRED SCENARIO
"WHAT DO YOU REALLY WANT?"
Helping Clients Set Goals That Make a Difference

Once clients understand either problem situations or opportunities for development more clearly, they often need help in determining what they want. They need to develop a preferred scenario, that is, a picture of a better future, choose specific goals to work on, and commit themselves to them. For instance, at this stage a troubled married couple is helped to outline what a better marital relationship might look like. Throughout the counseling process, rusty client imaginations need stimulating.

7

1. **"What do you want?" Help clients develop a range of possibilities for a better future.** If a client's current state of affairs is problematic and unacceptable, then he or she needs to be helped to conceptualize or envision a new state of affairs, that is, alternate, more acceptable possibilities. For instance, for a couple whose marriage is coming apart and who fight constantly, one of the elements of the new scenario might be fewer and fairer fights. Other possible elements of this better marriage might be greater mutual respect, more openness, more effectively managed conflicts, a more equitable distribution of household tasks, and so forth. Separation or even divorce might be considered if differences are irreconcilable and if the couple's values system permits such a solution.

2. **"What do you really want?" Help clients translate preferred-scenario possibilities into goals.** Once a variety of preferred-scenario possibilities — which constitute possible goals or desired outcomes of the helping process — have been generated, it is time to help clients choose the possibilities that make most sense for them and turn them into an agenda, that is, a goal or a "package" of goals to be accomplished. A new scenario is not a wild-eyed, idealistic state of affairs but rather a conceptualization or a picture of what the problem situation would be like if improvements were made. The agenda put together by the client needs to be viable, that is, capable of being translated into action. It is viable to the degree that it is stated in terms of clear and specific outcomes and is substantive or adequate, realistic, in keeping with the client's values, and capable of being accomplished within a reasonable time frame.

3. **"What are you willing to pay for what you want?" Help clients commit themselves to the goals they choose.** Problem-managing goals are useless if they are not actively pursued by clients. They need to determine what they are willing to pay for a better future and then search for incentives to help them move forward. The search for incentives is especially important when the choices are hard. How are truants with poor home situations to commit themselves to returning to school? What are the incentives for such a choice? Most clients, like most of the rest of us, struggle with commitment.

STAGE III: GETTING THERE
"WHAT DO YOU NEED TO DO TO GET WHAT YOU WANT?"
Helping Clients Formulate Strategies and Plans for Constructive Change

Discussing and evaluating preferred-scenario possibilities and choosing goals — the work of Stage II — determine the **"what"** of the helping process, that is, **what** must be accomplished by clients in order to manage their lives more effectively. Stage III deals with **how** goals are to be accomplished. Some clients know what they want to accomplish but need help in determining how to do it.

1. **Help clients brainstorm a range of strategies for accomplishing their goals.** In this step clients are helped to discover a number of different ways of achieving their goals. The principle is simple: Action strategies tend to be more effective when chosen from among a number of possibilities. Some clients, when they decide what they want, leap into action, implementing the first strategy that comes to mind. While such a bias toward action may be laudable, the strategy may be ineffective, inefficient, imprudent, or a combination of all three.

2. **Help clients choose action strategies that best fit their needs and resources.** If you do a good job in the first step of Stage III, that is, if you help clients identify a number of different ways of achieving their goals, then clients will face the task of choosing the best set. In this step your job is to help them choose the strategy or "package" of strategies that best fits their

preferences and resources. This tailoring of action strategies is important. One client might want to improve her interpersonal skills by taking a course at a college while another might prefer to work individually with a counselor while a third might want to work on his own.

3. **Help clients draw up a plan.** Once clients are helped to choose strategies that best fit their styles, resources, and environments, they need to assemble these strategies into a **plan**, a step-by-step process for accomplishing a goal. If a client has a number of goals, then the plan indicates the order in which they are to be pursued. Clients are more likely to act if they know what they need to do and in what order they should do it. Plans help clients develop discipline and also keep them from being overwhelmed by the work they need to do.

CLIENT ACTION:
THE HEART OF THE HELPING PROCESS
Helping Clients Act Both Within and
Outside Counseling Sessions

Helping is ultimately about problem-managing and opportunity-developing **action** that leads to outcomes that have a positive impact on clients' lives. Discussions, analysis, goal setting, strategy formulation, and planning all make sense only to the degree that they help clients to act prudently and accomplish problem-managing and opportunity-developing goals. There is nothing magic about change; it is hard work. If clients do not act on their own behalf, nothing happens.

Two kinds of client action are important here: First, actions within the counseling sessions themselves. The nine steps described above are not things that helpers do to clients, rather they are things that clients are helped to do. Clients must take ownership of the helping process. Second, clients must act "out there" in their real day-to-day worlds. Problem-managing and opportunity-developing action is ultimately the name of the game. The stages, steps, and action orientation of the helping process make sense to the degree that they lead to constructive change.

Since all the stages and steps of the helping process can be drivers of client action, action themes will be woven into each set of exercises. This will reinforce the principle that discussion and action go hand in hand.

CLIENT-FOCUSED HELPING

Helping usually does not take place in the neat, step-by-step fashion suggested by the stages and steps of the helping model. With some clients you will use only parts of the model just described. But you need the working knowledge and skill to be able to use whatever part is needed by the client. Effective helpers start wherever there is a client need. For instance, if a client needs support and challenge to commit himself or herself to realistic goals that have already been chosen, then the counselor tries to be helpful at this point. The nine steps of the helping model are active ways of **being with** clients in their attempts to manage problems in living and develop unused potential.

The needs of your clients and not the logic of a helping model should determine your interactions with them. One of the overriding needs of clients, of course, is to turn discussion and analysis of problem situations into problem-managing action. If you can help clients do this, you are worth your weight in gold.

9

SOME CAUTIONS IN USING THIS MANUAL

First, it is important to note that the exercises suggested in this manual are means, not ends in themselves. They are useful to the degree that they help you develop a working knowledge of the helping model and acquire the kinds of skills that will make you an effective helper. *They are helpful to the degree that they help you begin to become proficient in helping clients achieve the overall goals of the helping process.* Other exercises can be added, and the ones outlined here can be modified in order to achieve this goal more effectively.

Second, these exercises have been written as an adjunct to the text. They usually presuppose information in the text that is not repeated in the exercises. Also, the examples used in the text will help you do these exercises in a much more informed way.

Third, the length of training programs differs from setting to setting. In shorter programs there will not be time to do all these exercises. Nor is it necessary to do all of them. In shorter training programs, it is helpful to do at least one exercise from each of the sections on communication skills, one from each of the nine steps of the helping process, and one from the sections on action. This will help you develop a behavioral feeling for the **entire** helping model and not just the communication skills part of it. If the entire focus of the training program is on the communication skills that serve the helping process, you should understand that your training is incomplete. Further training in the helping model itself is needed.

Fourth, these exercises achieve their full effect only if you share them with a learning partner or the members of your training group and receive **feedback** on how well you are learning the model and the skills that make it work. Your instructor will set up the structure needed to do this. Since time limitations are always an issue, learning how to give brief, concise, behavioral feedback in a humane, caring way is most important.

Section 3
VALUES IN HELPING RELATIONSHIPS

It is important for you to take initiative in determining the kinds of values you want to pervade the helping process and relationship. Too often such values are afterthoughts. The position taken in Chapter Three of the text is that values should provide guidance for everything you do in your interactions with clients.

As indicated in the text, the values that are to permeate your helping relationships and practice must be owned by you. Learning about values from others is very important, but mindless adoption of the values promoted by the helping profession without reflection inhibits your owning and practicing them. You need to wrestle with your values a bit to make them your own. Second, your values must actually make a difference in your interactions with clients. That is, they must be values-in-use and not merely espoused values. Your clients will get a feeling for your values, not from what you say but from what you do.

EXERCISE 2: WHAT IF I WERE A CLIENT?

1. Picture yourself as a client. Think of some of the problems you have had to grapple with or

are struggling with now. Then picture yourself dealing with these with a counselor. Then ask yourself these questions. Jot down words, phrases, or simple sentences in response.

a. What would I want to get out of seeing a helper?

_____

_____

_____

_____

_____

b. What would I want the helper to be like?

_____

_____

_____

_____

_____

c. How would I want to be treated?

_____

_____

_____

_____

_____

2. Share your statements with a learning partner. Note both the similarities and the differences.
3. Finally, read Chapter Three in *The Skilled Helper*, Building the Helping Relationship: Values in Action. This chapter deals with values that should permeate the helping process together with a preliminary consideration of the relationship between client and helper. Compare the values implicit in your answers to the above questions with the values outlined in Chapter Three.

EXERCISE 3: YOUR PRELIMINARY STATEMENT OF VALUES

1. Write a preliminary STATEMENT OF VALUES based on your learnings from Exercise 1 that you as a helper could give to potential clients.

_____

_____

_____

_____

_____

_____

_____

_____

_____

2. Share your statement with a learning partner. Note both the similarities and the differences. Tell each other how you would feel in light of the other's STATEMENT OF VALUES, what your hopes and fears would be, were you to be your learning partner's client.

Section 4
ACTION ORIENTATION

Since the skilled-helper model is action-oriented, it is important for you to understand the place of "getting things done" in your own life. If you are to be a catalyst for problem-managing and opportunity-developing action on the part of clients, then reviewing your own track record in this regard is important. Unfortunately, helping often suffers from too much talking and not enough doing. Research shows that helpers are sometimes more interested in helping clients develop new insights than encouraging them to act on them. Inertia and procrastination plague most of us. The exercises in this section are designed to help you explore your own orientation toward action so that you may become a more effective stimulus to action for your clients. Read Chapter Four of _The Skilled Helper_, A Bias Toward Action.

EXERCISE 4: EXPLORING YOUR ACTION ORIENTATION

1. Read Chapter Four in the text. It deals with the kind of action orientation needed by helpers and clients alike and discusses some of the blocks to problem-managing action.

In the following example, a counselor trainee discusses ways in which she procrastinates and outlines a project she has been putting off.

Example: Dahlia, 55, whose children are now grown, has returned to school in order to become a counselor. She has this to say about her action orientation.

Chronic Procrastination Scenario. "My husband is in business for himself. I take care of a lot of the routine correspondence for the business and our household. Often I let it pile up. The more it piles up, the more I hate to face it. On occasion, an important business letter gets lost in the shuffle. This annoys my husband a great deal. Then with a great deal of flurry, I do it all and for a while keep current. But then I slide back into my old ways. I also notice that when I let it pile up I waste a lot of time reading junk mail — catalogs of things I'm not going to buy, things like that."

A Project Being Put Off. "I have an older sister who is a widow. She has one autistic child, nearly 20 now, who is at home. He goes to a school for autistic citizens — it's not just kids — and is gone several hours every week day. Lately, he is becoming more difficult to manage. He has temper outbursts and things like that. This puts a great deal of stress on my sister. She's much more timid than I am. The day will come when she can no longer take care of him. I have told her that I would help her in the whole process of placing him in an institution of some kind. I know she can use my help. If I don't do anything, things will just get worse. One of my major concerns is ending up doing something *for* my sister. I have to do whatever I'm going to do *with* her. I've been putting off doing anything about it because I know it will be extremely difficult for her. It's not my favorite project. I think about it at least once a day, then I put it out of my mind."

2. Like Dahlia, identify one or more ways in which you chronically put off action.

_____

_____

_____

_____

_____

3. Like Dahlia, describe a project that you have been putting off.

_____

_____

_____

13

_____

_____

_____

4. Get together with a learning partner and tell each other what you learned about yourself and your "bias toward action" from doing steps 2 and 3.

_____

EXERCISE 5: EXPLORING THE SELF-STARTER IN YOURSELF

While all of us have a tendency to put things off, we also have the possibility of become self-starters. Self-starters move to problem-managing and opportunity-developing action without being influenced, asked, or ordered to do so. Skilled helpers are, ideally, self-starters who help clients explore self-starting possibilities in themselves. Read the following example.

Example: Dahlia, in searching for the self-starter within herself, discovered the following:

- I never put off the things I like to do. For instance, I like the volunteer work I do at the hospital. No one has to put pressure on me to show up. I even volunteer for added special projects. I also plan holidays for the family very carefully. Planning them is half the fun.
- I spent a week at a resort one year that had no television. I did all sorts of creative things with the extra time I had. I finished reading a biography that I couldn't find the time to read back home. I learned that I'm a much better self-starter when I get rid of some of the clutter. Television for me is part of the clutter.
- Once I'm into a task I've been putting off, I keep going. Just starting is the hard part. When I keep putting something off, I'm miserable. I'm beginning to learn that "JUST DO IT!" should be my motto.

1. Describe the ways or the areas of life in which you are a self-starter.

_____

_____

_____

_____

_____

2. Share what you have learned about yourself with a learning partner. See if some of your partner's learnings apply to you.

BASIC COMMUNICATION SKILLS FOR HELPERS

There are two sets of communication skills that are essential for helpers. The first set includes attending, listening, basic empathy, and probing and is the focus of the exercises in Part Two. The second set, communication skills associated with helping clients challenge themselves, is dealt with in the exercises associated with Step I-B of the helping model — helping clients challenge their blind spots and develop new, action-oriented perspectives.

In Part II, Section 5 deals with attending and listening, while Section 6 focuses on the communication of accurate empathic understanding and probing.

Section 5
COMMUNICATION SKILLS I: ATTENDING AND LISTENING

Much of helping takes place through a dialogue between client and helper. If this dialogue is to serve the overall problem-management and opportunity-development goals of the helping process, the quality of the dialogue is critical. In this section the exercises focus on (I) your "attending" behavior and (II) active listening.

I. EXERCISES IN ATTENDING

Your posture, gestures, facial expressions, and voice all send nonverbal messages to your clients. The purpose of the exercises in this section is to make you aware of the different kinds of nonverbal messages you send to clients through such things as body posture, facial expressions, and voice quality and how to use nonverbal behavior to communicate with them. It is important that what you say verbally is reinforced rather than muddled or contradicted by your nonverbal messages. Before doing these exercises, read Chapter Five in *The Skilled Helper*, Communication Skills I: Attending and Listening. There are two points. First, use your posture, gestures, facial expressions, and voice to *send messages* you want to clients, such as, "I want to work with you to help you manage your life better." Second, attend carefully so that you can *listen* carefully to clients.

EXERCISE 6: BECOMING AWARE OF YOUR ATTENDING BEHAVIORS IN EVERDAY LIFE

This is an exercise you do outside the training group in your everyday life. Observe your attending behaviors for a week — at home, with friends, at school, at work. You are not being asked to become preoccupied with the micro-behaviors of attending. Observe the quality of your presence to others when you engage in conversations with them. Of course, even being asked to "watch yourself" will induce changes in your behavior; you will probably use more effective attending behaviors than you ordinarily do. The purpose of this exercise is to sensitize you to attending behaviors in general and to get some idea of what your day-to-day attending style looks like. First, read about attending skills in the text.

1. Read the parts of Chapter Five, Attending and Listening, that deal with attending behavior.
2. Watch yourself for a week attending (or not attending) to others in various social settings.
3. What are you like when you are with and listening to others, especially in a serious situation? What do you do well? What needs improvement in your attending style?

_____

_____

_____

_____

_____

_____

EXERCISE 7: OBSERVING AND GIVING FEEDBACK ON QUALITY OF PRESENCE

In the training sessions, make sure that your nonverbal behavior is helping you work effectively with others and sending the messages you want to send. Throughout the training program, observe the nonverbal behavior of your fellow trainees and give them feedback on how it affects you when you play the role of client or observer. Throughout the training program, ask for feedback on your own attending style.

Exercise 7, then, is an exercise that pertains to the entire length of the training program. You are asked to give ongoing feedback to yourself and to the other members of the training group on the quality of your presence to one another as you interact, learn, and practice helping skills. Recall especially the basic elements of physical attending summarized by the acronym *SOLER*:

S   Face your clients **SQUARELY**. This says that you are available to work with them.
O   Adopt an **OPEN** posture. This says that you are open to your clients and want to be nondefensive.

L   **LEAN** toward the client at times. This underscores your attentiveness and lets clients know that you are with them.

E   Maintain good **EYE** contact without staring. This tells your clients of your interest in them and their concerns.

R   Remain relatively **RELAXED** with clients as you interact with them. This indicates your confidence in what you are doing and also helps clients relax.

Of course, these are guidelines rather than hard and fast rules. Here is a checklist to help you provide that feedback to your fellow helpers.

- How effectively is the helper using postural cues to indicate a willingness to work with the client?
- In what ways does the helper distract clients and observers from the task at hand, for instance, by fidgeting?
- How flexible is the helper when engaging in *SOLER* behaviors? To what degree do these behaviors help the counselor be with the client effectively?
- How natural is the helper in attending to the client? Are there any indications that the helper is not being himself or herself?
- From an attending point of view, what does the helper need to do to become more effectively present to his or her clients?

More important than nonverbal behavior in itself is the total quality of your being with and working with your clients. Your posture and nonverbal behavior are a part of your presence, but there is more to presence than *SOLER* activities. Since quality of presence involves both internal attitudes and external behaviors, trainees should not become preoccupied with the micro-skills of attending.

II. EXERCISES IN ACTIVE LISTENING

Read the sections on listening in Chapter Five of *The Skilled Helper*. Effective helpers are active listeners. When you listen to clients, you listen to them discussing:

- their **experiences**, what they see as happening *to* them;
- their **behaviors**, what they do or fail to do;
- their **affect**, the feelings and emotions that arise from their experiences and behaviors; and
- their **points of view** in talking about their experiences, behaviors, and feelings.

Helpers listen carefully in order to be able to respond both with understanding and to help clients challenge themselves. Clients' experiences, behaviors, feelings, and points of view constitute the *building blocks* of both empathic understanding and challenging.

Experiences, behaviors, and feelings can be either overt (capable of being seen by others) or covert (not seen by others, hidden "inside" the speaker).

- **Overt experience:** "He yelled at me."
- **Covert experience:** "Thoughts about death come out of nowhere and flood my mind."

- **Overt behavior:** "I spend about three hours every night in some bar."
- **Covert behavior:** "Before she comes over I plan everything I'm going to say."

- **Overt emotion** (expressed): "I got very angry and shouted at her."
- **Covert emotion:** (felt, but not expressed): "I was delighted that he failed, but I didn't let on."

You can learn a great deal about clients by listening to the mix of experiences, behaviors, and feelings they discuss and how specific or vague they are.

EXERCISE 8: LISTENING TO YOUR OWN FEELINGS AND EMOTIONS

If you are to listen to the feelings and emotions of clients, you first should be familiar with your own emotional states. A number of emotional states are listed below. You are asked to describe what you feel when you feel these emotions. Describe what you feel as *concretely* as possible: How does your body react? What happens inside you? What do you feel like doing? Consider the following examples.

Example 1 — Accepted:

When I feel accepted,

- I feel warm inside.
- I feel safe.
- I feel free to be myself.
- I feel like sitting back and relaxing.
- I feel I can let my guard down.
- I feel like sharing myself.
- I feel some of my fears easing away.
- I feel at home.
- I feel at peace.
- I feel my loneliness drifting away.

Example 2 — Scared:

When I feel scared,

- my mouth dries up.
- my bowels become loose.
- there are butterflies in my stomach.
- I feel like running away.
- I feel very uncomfortable.
- I feel the need to talk to someone.
- I turn in on myself.
- I'm unable to concentrate.
- I feel very vulnerable.
- I sometimes feel like crying.

1. Choose four of the emotions listed below or others not on the list. Try your hand at the emotions you have difficulty with. It's important to listen to yourself when you are experiencing emotions that are not easy for you to handle.
2. Picture to yourself situations in which you have actually experienced each of these emotions.
3. Then, as in the example above, write down in detail what you experienced.

1. accepted
2. affectionate
3. afraid
4. angry
5. anxious
6. attracted
7. bored

8. competitive
9. confused
10. defensive
11. desperate
12. disappointed
13. free
14. frustrated

15. guilty
16. hopeful
17. hurt
18. inferior
19. interested
20. intimate
21. jealous

22. joyful	28. satisfied
23. lonely	29. shocked
24. loving	30. shy
25. rejected	31. superior
26. respected	32. suspicious
27. sad	33. trusting

The reason for this exercise is to sensitize yourself to the wide variety of ways in which clients express and name their feelings and emotions.

LISTENING TO EXPERIENCES AND BEHAVIORS

Although the feelings and emotions of clients (not to mention your own) are extremely important, sometimes helpers concentrate too much, or rather too exclusively, on them. Feelings and emotions need to be understood, both by helpers and by clients, in the **context** of the experiences and behaviors that give rise to them. On the other hand, when clients hide their feelings, both from themselves and from others, then it is necessary to listen carefully to cues indicating the existence of suppressed, ignored, or unmanaged emotion.

EXERCISE 9: LISTENING TO KEY EXPERIENCES, BEHAVIORS, AND FEELINGS

Since feelings and emotions do not arise in a vacuum, in this exercise you are asked to "listen to" and identify the kinds of experiences and behaviors that give rise to the client's feelings. In the following cases you are asked to:

1. Listen very carefully to what the client is saying.
2. Identify the client's key experiences, what he or she says is happening to him or her.
3. Identify the client's key behaviors, what he or she is doing or not doing.
4. Identify the key feelings and emotions associated with these experiences and behaviors.

Example: A twenty-seven-year-old man is talking to a minister about a visit with his mother the previous day. "I just don't know what got into me! She kept nagging me the way she always does, asking me why I don't visit her more often. As she went on, I got more and more angry. (He looks away from the counselor down toward the floor.) I finally began screaming at her. I told her to get off my case. (He puts his hands over his face.) I can't believe what I did! I called her a bitch. (Shaking his head.) I called her a bitch about ten times, and then I left and slammed the door in her face."

a. **Key experiences:** Mother's nagging.
b. **Key behaviors:** Losing his temper with his mother, yelling at her, calling her a name, slamming the door in her face.
c. **Feelings/emotions generated:** He feels embarrassed, guilty, ashamed, distraught, extremely disappointed with himself, remorseful.

Note carefully: This man is talking **about** his anger, the way he let his temper get away from him, but while talking to the minister, he is feeling and expressing the emotions listed above.

1. A woman, 40, married, no children, who has had several sessions with a counselor. She went because she was bored and felt that all the "color" had gone out of her life: "These counseling sessions have really done me a great deal of good! I've worked hard in these sessions, and it's paid off. I enjoy my work more. I actually look forward to meeting new people. My husband and I are talking more seriously and decently to each other. At times he's even tender toward me the way he used to be. Now that I've begun to take charge of myself more and more, there's just so much more freedom in my life!"

a. **Client's key experiences:**

_____

_____

b. **Client's key behaviors:**

_____

_____

c. **What feelings/emotions do these experiences and behaviors generate?**

_____

_____

2. A man, 64, who has been told that he has terminal cancer, speaking to a medical resident: "Why me? Why me? I'm not even that old! I keep looking for answers, and there are none. I've sat for hours in church, and I come away feeling empty. Why me? I don't smoke or anything like that. (He begins to cry.) Look at me. I thought I had some guts. I'm just a slobbering mess. Oh God, why terminal? What are these next months going to be like? (Pause, he stops crying.) Why would you care! I'm just a failure to you guys."

a. **Client's key experiences:**

_____

_____

b. **Client's key behaviors:**

_____

_____

c. **What feelings/emotions do these experiences and behaviors generate?**

_____

_____

3. A woman, 38, unmarried, talking about losing a friend: "My best friend has just turned her back on me. And I don't even know why! (said with great emphasis) From the way she acted, I think she has the idea that I've been talking behind her back. I simply have not! (also said with great emphasis) Damn! This neighborhood is full of spiteful gossips. She should know that. If she's been listening to those foulmouths who just want to stir up trouble. . . . She could at least tell me what's going on."

a. **Client's key experiences:**

_____

_____

b. **Client's key behaviors:**

_____

_____

c. **What feelings/emotions do these experiences and behaviors generate?**

_____

_____

4. A man, 54, talking to a counselor about a situation at work: "I don't know where to turn. They're asking me to do things at work that I just don't think are right. If I don't do them, well, I'll probably be let go. And I don't know where I'm going to get another job at my age in this economy. But if I do what they want me to, I think I could get into trouble, I mean legal trouble. I'd be the fall guy. My head's spinning. I've never had to face anything like this before. Where do I turn?"

a. **Client's key experiences:**

_____

_____

b. **Client's key behaviors:**

_____

_____

c. **What feelings/emotions do these experiences and behaviors generate?**

_____

_____

5. A girl in her late teens who is serving a two-year term in a reformatory speaks to a probation counselor: (She sits silently for a while and doesn't answer any question the counselor puts to her. Then she shakes her head and looks around the room.) "I don't know what I'm doing here. You're the third counselor they've sent me to . . . or is it the fourth? It's a waste of time! Why do they keep making me come here? (She looks straight at the counselor.) Let's fold the show right now. You're not getting anything out of me. Come on, get smart."

a. **Client's key experiences:**

_____

_____

b. **Client's key behaviors:**

_____

_____

c. **What feelings/emotions do these experiences and behaviors generate?**

_____

_____

Supplemental exercises are found in Appendix Two.

EXERCISE 10: LISTENING TO THE CLIENT'S POINT OF VIEW

Empathic listening involves listening to and understanding the client's point of view in terms of experiences, behaviors, and feelings. Even when you think the client's point of view needs to be challenged, it is essential to hear it. The following instructions apply to all three of the following client summaries.

1. Read the paragraph. Try to picture the clients saying what they say. Listen carefully.
2. Go over the paragraph sentence by sentence. Identify experiences, behaviors, and feelings.
3. Summarize the client's point of view in terms of key experiences, key behaviors, and key feelings and emotions. Do not evaluate it or contaminate it with your own point of view.

1. The following client is a 40-year-old woman who has just lost her job. She is talking about the events before, at the time, and after she was fired. "Yesterday I was talking with one of the punch-press operators when my boss storms in and begins raking me over the coals for a work stoppage I had nothing to do with. I stood there in shock. I was so angry that I wanted to yell back at him, but I kept my cool. But all day I couldn't get it out of my mind. No matter what I was doing, it haunted me. I finally got so angry that I burst into his office and told him just what I thought of him. I even let him have it for a few lousy things he's done in the past. He fired me on the spot. Last night I was pretty depressed. And all day today I've been trying to figure out where I can get a new job or maybe how I can get my old job back."

What is this client's point of view?

_____

_____

_____

_____

_____

2. This client is a 37-year-old man who is talking to a counselor for the first time. He has been referred by a doctor who has found no physical basis for a variety of somatic complaints. "My wife keeps putting me down. For instance, last week she got a job without even discussing it with me. She didn't even ask me how I'd feel. She doesn't share what's going on inside. She makes big decisions without letting me in on the process. I'm sure she sees me as weak and ineffectual. She's just like her mother. My mother-in-law never wanted me to marry her. She was too good for me. Now my wife does everything to prove that her mother is right. She would never admit it, of course, but that's the way she is. I wouldn't be surprised if her goal is to earn more money than I do. I see other guys getting divorces for a lot less than I have to put up with. But that would make both of them happy. I asked her to come with me to see you, and she laughed at me, she actually laughed at me."

What is this client's point of view?

_____

_____

_____

_____

_____

3. The following is a 24-year-old gay male who recently learned that he had tested postive for the HIV virus. He has received counseling to help him deal with the shock through the clinic where he was tested. Now, about two months later, he is talking to a different counselor. "I'm sort of over the initial shock, that is, if you're ever over the shock. But now I'm trying to put my life back together again. I'm finding that my desire for sex is not any less than it was before. I got the virus because I was a bit too promiscuous and not careful enough. But I want to be with people again. Being gay, I know, means being more promiscuous that if I were straight. I can't seem to help it. It's the way I am — the way we are. Anyway, being tied down to one person, well, who'd have me now? I've got to find a way of straightening this all out."

What is this client's point of view?

_____

_____

_____

_____

_____

_____

Share your summaries of each of these client's point of view with your learning partner. First discuss just the client's point of view. Then discuss what you believe needs to be challenged in this point of view.

EXERCISE 11: COMMUNICATING UNDERSTANDING OF ONE ANOTHER'S POINTS OF VIEW

Another person's point of view is made up not only of experiences, behaviors, and feelings but also the interpretation or the slant the person gives to these. Empathy, then, includes this slant, even though you think the slant needs to be challenged. Challenging a person's interpretations of his or her experiences, behaviors, and feelings may be necessary. This is dealt with in Step I-B of the helping model.

1. Divide into groups of three. The roles in each group are speaker, listener, and observer.
2. Take a few minutes to prepare a statement on an issue that you believe to be important. You may jot down a few notes, but the statement is to be spoken, not read. You can provide a bit of context, but the statement should be relatively short.
3. After determining who is to go first, the speaker delivers his or her statement to the listener, while the observer watches.
4. The listener listens carefully and then summarizes the speaker's point of view in the third person. The listener begins with the phrase: "This, I believe, is the speaker's point of view." The listener's summary should be relatively brief and deal with the speaker's core message or messages.

5. The observer gives feedback to the listener on his or her conciseness and accuracy.

6. The process continues until each member of the group has played all three roles.

Example: Janine is the first to take the speaker's role. She says: "As you can tell, I have a speech defect. As you might not be able to tell, at least not immediately, I'm also fairly bright. Also, while I'm not a stunning beauty, I'm not that bad looking. But the first thing a lot of people latch onto is the speech problem. I have more than a sneaking suspicion that this colors their view of me. My looks and my intelligence are seen through the filter of my speech. Often enough, I get discounted. I'm not exactly blaming people for that. It's so easy to do. But it leaves me feeling defensive much of the time. And I find that very uncomfortable."

Bernice, in the listener's role, summarizes Janine's point of view like this: "This, I believe, is Janine's point of view. She's angry at people because they don't take her as she is. She'd like to tell them about herself, but they don't want to listen. She has to defend herself all the time, and that's annoying."

Carla, in the observer role, gives feedback to Bernice: "I think the main message is correct: Janine wants to be taken for the full person she is despite the speech problem. But I heard disappointment rather than anger from Janine. Also she said that she feels defensive much of the time. She didn't say that she goes around defending herself."

Section 6
COMMUNICATION SKILLS II: BASIC EMPATHY AND PROBING

The payoff of attending and listening lies in the ability to communicate to clients an understanding of their experiences and behaviors and the feelings and emotions they generate. Furthermore, listening to clients' points of view enables you to let them know that you see their point of view even when you think that this point of view needs to be challenged or transcended. There are two parts to this section. Part I focuses on empathy; Part II adds the skill of probing.

I. COMMUNICATING UNDERSTANDING: EXERCISES IN BASIC EMPATHY

Basic empathy is the communication to another person of your understanding of his or her point of view with respect to his or her experiences, behaviors, and feelings. It is a skill you need throughout the helping model. Focusing on the client's point of view without necessarily agreeing with it is very useful in establishing and developing relationships with clients and in helping them clarify both problem situations and unexploited opportunities, in setting goals, and in developing strategies and plans. The starting point of the entire helping process and each of its steps is the client's point of view, even when it needs to be challenged.

The exercises in the previous section emphasized your ability to listen to and understand the client's point of view. The exercises in this section relate to your ability to **communicate** this understanding to the client.

EXERCISE 12: COMMUNICATING UNDERSTANDING OF A CLIENT'S FEELINGS

When feelings and emotions do constitute a part of a client's core message, an understanding of them needs to be communicated to him or her. Clients express feelings, and helpers can communicate an understanding of feelings in a variety of ways:

- **By single words:** I feel good. I'm depressed. I feel abandoned. I'm delighted. I feel trapped. I'm angry.
- **By different kinds of phrases:** I'm sitting on top of the world. I feel down in the dumps. I feel left in the lurch. I feel tip top. My back's up against the wall. I'm really steaming.
- **By what is implied in a behavioral statement** (what action I feel like taking): I feel like giving up (implied emotion: despair). I feel like hugging you (implied emotion: joy). I feel like smashing him in the face (implied emotion: severe anger). Now that it's over, I feel like throwing up (implied emotion: disgust).
- **By what is implied in experiences that are revealed:** I feel I'm being dumped on (implied feeling: anger). I feel I'm being stereotyped (implied feeling: resentment). I feel I'm at the top of her list (implied feeling: elation). I feel I'm going to catch my lunch (implied feeling: apprehension). Note that the implication of each could be spelled out: I feel angry because I'm being dumped on. I resent the fact that I'm being stereotyped. I feel great because I believe I'm at the top of her list. I'm apprehensive because I think I'm going to catch my lunch.

1. A number of situations involving different kinds of feelings and emotions are listed below. Picture yourself talking to this person.
2. Use two of the four ways of communicating understanding of the client's feelings listed below.

Example: Sally tells you that she has just been given the kind of job she has been looking for for the past two years.

- **Single word:** You're really happy.
- **A phrase:** You're on cloud nine.
- **Experiential statement:** You feel you got what you deserve.
- **Behavioral statement:** You feel like going out and celebrating.

Now express the following feelings and emotions in two different ways (single word, phrase, experiential statement, and/or behavioral statement).

1. This woman is about to go to her daughter's graduation from college: "I never thought that my daughter would make it through. I've invested a lot of money in her education. But more to the point, a lot of emotion. Now the day has arrived!"

_____

_____

_____

_____

2. This woman has just had her purse stolen. She's talking to a policewoman. "I had just cashed my bi-weekly paycheck, and the money was in the purse. I've had a streak of bad luck. My sister

was in an auto accident last week. And my only son was detained by the police for a minor theft earlier this week. There has not been much good news at all."

_____

_____

_____

_____

3. This man is waiting for the results of medical tests. He is talking to a hospital volunteer: "I've been losing weight for about two months and feeling tired and listless all the time. I'm afraid of what these tests are going to say. I've been putting them off. Well, now the waiting's getting to me. I . . . well, I just don't know where I stand. Nobody said anything to me during the tests."

_____

_____

_____

_____

_____

4. A prospective employer has just found out that this client has a criminal record: "I had hoped that I would get the job and prove myself before anyone found out about my record. I guess I was just stupid. I've just received a call from him telling me that I'm no longer being considered for the job. Well, I did what I thought was right. I never had the intention of deceiving anyone. I'm going to make it, somehow."

_____

_____

_____

_____

5. This woman has just lost a custody case for her children (during the interview she seems almost in a daze): "I never dreamed that the court would award custody to my husband. He's so selfish and spiteful. It's all over now."

_____

_____

3. Compare your responses with those of your learning partner. Then see if the two of you can improve upon your combined responses.

Supplemental exercises are found in Appendix Two.

EXERCISE 13: USING A FORMULA TO EXPRESS EMPATHY

Empathy focuses on the client's key messages — key experiences and/or key behaviors plus the feelings and emotions they generate. In this exercise you are asked to use the formula explained in the text: "You feel . . ." (followed by the right emotion and some indication of its intensity) "because. . ." (followed by the key experiences and/or behaviors that give rise to the emotion).

Example: A woman in a self-help group is talking about a relationship with a man. She says: "He began being abusive, calling me names, describing my defects. Even though I've talked about this in this group, I sat there very passively and just took it. I just took it!" A group member replies: "You feel very annoyed with yourself because you let him get away with it again even though you knew better."

1. First-year college student talking to counselor in the Center for Student Services: "Jane and Sue showed up at the party in dresses *and* with dates. And there I was, alone and in slacks! I mean, just typical of me!"

_____

_____

2. Man, 65, talking to a mental health counselor: "My wife died last year, and this year my youngest son went away to college. The other children are married. So now that I'm retired, I spend a lot of time rambling around a house that's . . . (pauses, looks out the window for a while) . . . really too big for me."

_____

_____

3. Married woman, 33, having a "solo" session with a marriage counselor: "I can't believe it! Tom came home on time for supper every day last week. I never thought he would live up to the contract we made. And yet, there he was, each and every night."

_____

4. Man, 40, talking about his invalid mother to a minister at his church: "She uses her illness to control me. It's a pattern; she's been controlling me all her life. (He grits his teeth and sets his jaw.) I bet she'll even make me feel responsible for her death."

_____

_____

5. Young woman, 25, talking about her current boyfriend to an older confidante: "I can't quite figure him out. I still can't tell if he really cares about me, or if he's just trying to get me into bed. It leaves everything up in the air."

_____

_____

Supplemental exercises are found in Appendix Two.

_____

EXERCISE 14: USING YOUR OWN WORDS TO EXPRESS EMPATHY

In this exercise you are asked to do two things:

1. Use the "you feel . . . because . . ." formula to communicate empathy to the client.
2. Recast your response in your own words while still identifying both feelings and the experiences and/or behaviors that underlie the feelings. But avoid advice giving.

Example: A married woman, 31, is talking to a counselor about her marriage: "I can't believe it! You know when Tom and I were here last week we made a contract that said that he would be home for supper every evening and on time. Well, he came home on time every day this past week. I never dreamed that he would live up to his part of the bargain so completely!"
• **Formula.** "You feel great because he really stuck to his word!"
• **Non-formula.** "He really surprised you by doing it right!"

Now imagine yourself listening intently to each of the following clients. First use the "You feel . . . because . . ." formula; then use your own words. Try to make the second response sound as natural (as much like yourself) as possible. After you use your own words, check to see if you have both a "you feel" part and a "because" part in your response.

1. A man, 40, is talking about his invalid mother: "I know she's using her illness to control me. How could a 'good' son refuse any of her requests at a time like this? (He pounds his fist on the arm of his chair.) But it's all part of a pattern. She's used one thing or another to control me all my life. If I let things go on like this, she'll make me feel responsible for her death!"

a. Use the formula.

_____

_____

b. Use your own words.

_____

_____

2. A woman, 25, talking about her current boyfriend: "I can't quite figure him out. (She pauses, shakes her head slowly, and then speaks quite slowly.) I just can't figure out whether he really cares about me or if he's just trying to get me into bed. I've been burned before; I don't want to get burned again."

a. Use the formula.

_____

_____

b. Use your own words.

_____

_____

3. A businessman, 38, talking to a company counselor: "I really don't know what my boss wants. I don't know what he thinks of me. He tells me I'm doing fine even though I don't think that I'm doing anything special. Then he blows up over nothing at all. I keep asking myself if there's something wrong with me, I mean, that I don't see what's getting him to act the way he does. I'm beginning to wonder if this is the right job for me."

a. Use the formula.

_____

_____

b. Use your own words.

_____

_____

4. A woman, 73, in the hospital with a broken hip: "When you get old, you have to expect things like this to happen. It could have been much worse. When I lie here, I keep thinking of the people in the world who are a lot worse off than I am. I'm not a complainer. Oh, I'm not saying that this is fun or that the people in this place give you the best service — who does these days? — but it's a good thing that these hospitals exist. Think of those who don't have anything."

a. Use the formula.

_____

_____

b. Use your own words.

_____

_____

5. A seventh-grade girl to her teacher, outside class: "My classmates don't like me, and right now I don't like them! Why do they have to be so mean? They make fun of me — well, they make fun of my clothes. My family can't afford what some of those dopes wear. Gee, they don't have to like me, but I wish they'd stop making fun of me."

a. Use the formula.

_____

_____

b. Use your own words.

_____

_____

Supplemental exercises are found in Appendix Two.

EXERCISE 15: EMPATHY WITH CLIENTS FACING DILEMMAS

Clients sometimes talk about conflicting values, experiences, behaviors, and emotions. Responding with empathy means communicating an understanding of the conflict. Consider the following example.

Example: A woman, 32, talking to a counselor about adopting a child: "I'm going back and forth, back and forth. I say to myself, 'I really want a child,' but then I think about Bill [her husband] and his reluctance. He so wants our own child and is so reluctant to raise someone else's. We don't even know why we can't have children. But the fertility specialists don't offer us much hope. At times when I so want to be a mother I think I should marry someone who would be willing to adopt a child. But I love Bill and don't want to point an accusing finger at him."

- **Identify the conflict or dilemma.** She believes that she runs the risk of alienating her husband if she insists on adopting a child, even though she strongly favors adoption.
- **Formula.** "You feel trapped between your desire to be a mother and your love for your husband."
- **Non-formula.** "You're caught in the middle. Adopting a child would solve one problem but perhaps create another."

1. A factory worker, 30: "Work is okay. I do make a good living, and both my family and I like the money. My wife and I are both from poor homes, and we're living much better than we did when we were growing up. But the work I do is the same thing day after day. I may not be the world's brightest person, but there's a lot more to me than I use on those machines."

a. The conflict.

_____

b. Use the formula.

_____

_____

c. Use your own words.

_____

_____

_____

2. A mental hospital patient, 54, who has spent five years in the hospital; he is talking to the members of an ongoing therapy group. Some of the members have been asking him what he's doing to get out. They point out the tendency to "push people out." He says, "To tell the truth, I like it here. I'm safe and secure. So why are so many people here so damn eager to see me out? Is it a crime because I feel comfortable here? (Pause, then in a more conciliatory voice.) I know you're all interested in me. I see that you care. But do I have to please you by doing something I don't want to do?"

a. The conflict.

_____

b. Use the formula.

_____

_____

c. Use your own words.

_____

_____

_____

3. A juvenile probation officer to a colleague: "These kids drive me up the wall. Sometimes I think I'm really stupid for doing this kind of work. They taunt me. They push me as far as they can. To some of them I'm just another 'pig.' But every time I think of quitting — and this gets me — I know I'd miss the work and even miss the kids one way or another. When I wake up in the morning, I know the day's going to be full and it's going to demand everything I've got."

a. The conflict.

_____

b. Use the formula.

_____

_____

c. Use your own words.

_____

_____

_____

EXERCISE 16: THE PRACTICE OF EMPATHY IN EVERYDAY LIFE

If the communication of accurate empathy is to become a part of your natural communication style, you will have to practice it outside formal training sessions. That is, it must become part of your everyday communication style or it will tend to lack genuineness in helping situations. Practicing empathy "out there" is a relatively simple process.

1. **Empathy as an improbable event.** Empathy is not a normative response in everyday conversations. Find this out for yourself. Observe everyday conversations. Count how many times empathy is used as a response in any given conversation.

2. **Your own use of empathy.** Next try to observe how often you use empathy as part of your normal style. In the beginning, don't try to increase the number of times you use empathy in day-to-day conversations. Merely observe your usual behavior. What part does empathy normally play in your style?

3. **Increasing your empathic responses.** Begin to increase the number of times you use accurate empathy. Be as natural as possible. Do not overwhelm others with this response; rather, try to incorporate it gradually into your style. You will probably discover that there are quite a few opportunities for using empathy without being phony. Keep some sort of record of how often you use empathy in any given conversation.

4. **The impact of empathy.** Observe the impact your use of empathy has on others. Don't set out to use others for the purpose of experimentation. But, as you gradually increase your use of this communication skill naturally, try to see how it influences your conversations. What impact does it have on you? What impact does it have on others?

5. **Learnings.** In a forum set up by the instructor, discuss with your fellow trainees what you have learned from this "experiment."

If empathy becomes part of your communication style "out there," then you should appear more and more natural in using empathy in the training program, both in playing the role of the helper and in listening and providing feedback to your fellow trainees. On the other hand, if you use empathy only in the training sessions, it will most likely remain artificial.

II. EXERCISES IN THE USE OF PROBES

It is not enough to help clients choose problems, issues, or concerns that make a difference. Once a problem situation is chosen, the specific issues within it need to be identified. If Connie and Chuck want to deal with the poor communication they have with each other in their marriage, then the key issues relating to communication or the lack thereof need to be identified and explored. You do this by using empathy, probing, and challenge to achieve the kind of problem clarity that can lead to goal setting and strategy formulation. A problem or unused opportunity is clear if it is spelled out in terms of specific experiences, specific behaviors, and specific feelings in specific situations. To do the next four exercises well, you are asked to use probes with yourself — probes such as "Could you make that more specific? What else? What do you mean by that?" Once you appreciate what it takes to talk concretely and specifically about your own issues, then it will be easier for you to use probes to help clients do the same.

EXERCISE 17: SELF-PROBING — SPEAKING CONCRETELY ABOUT YOUR EXPERIENCES

Clients often speak too vaguely about their problems. Helpers often go "round the mulberry bush" with them, allowing them to be too general. Vague problems lead only to vague solutions that tend to be worthless. Since problem situations need to be spelled out in terms of experiences, behaviors, and feelings, we start with experiences, your experiences. In this exercise, you are asked to speak of some of your experiences, first vaguely, then concretely; that is, you are asked to use probes such as Who? What? When? Where? How? to force yourself to speak more specifically.

Example 1: George, a counselor trainee, discusses his "inefficiency."

- **Vague statement of experience.** "I'm sometimes less efficient than I could be because of headaches."
- **Concrete statement of the same experience.** "I get migraine headaches about once a week. They make me extremely sensitive to light and usually cause severe pain. I often get so sick that I throw up. They happen more often when I'm tense or under a lot of pressure. For instance, I often come away from a visit with my ex-wife with one. Each week they rob me of productive hours of work, either at work or at home."

Example 2: Jane, a counselor trainee, discusses her marriage.

- **Vague statement of experience.** "My marriage is falling apart."
- **Concrete statement of the same experience.** "My husband is going around with other women, though he won't admit it. He never asks me to have sex, though occasionally it 'happens.' He is verbally abusive at times, though he has never hit me. I keep most of this to myself."

1. In the spaces following, explore three experiences, things you see as happening *to* you, that are related to some problem situation or situations of your own. Choose issues that might affect the quality of your helping.

a. **Vague.** _____

_____

b. **Concrete.** _____

_____

_____

_____

a. **Vague.** _____

_____

b. **Concrete.** _____

_____

_____

a. **Vague.** _____

_____

b. **Concrete.** _____

_____

_____

_____

_____

2. Share one or two of these with a learning partner. Get feedback on how clear your statement is. If you do not think that your partner's statement is as clear as it might be, use probes to help him or her make the statement clearer.

EXERCISE 18: SELF-PROBING — SPEAKING CONCRETELY ABOUT YOUR BEHAVIOR

In this exercise, you are asked to speak about some of your behaviors (what you do or fail to do) that are involved in some problem situation. As in the preceding exercise, start with a vague statement, then clarify it with the kind of detail needed to serve the problem-management process. Choose a problem situation that might affect you in your role as helper.

Example 1: Karen writes about her tendency to dominate.

- **Vague statement of behavior.** "I tend to be domineering."
- **Concrete statement of the same behavior.** "I try, usually in subtle ways, to get others to do what I want to do. I even pride myself on this. In conversations, I take the lead. I interrupt others, jokingly and in a good-natured way, but I make my points. If a friend is talking about something serious when I'm not in the mood to hear it, I change the subject."

Example 2: Eric discusses his relationship with his wife.

- **Vague statement of behavior.** "I don't treat my wife right."
- **Concrete statement of the same behavior.** "When I come home from work, I read the paper and watch some TV. I don't talk much to my wife except a bit at supper. I don't share the little things that went on in my day. Neither do I encourage her to talk about what happened to her. Still, if I feel like having sex later, I expect her to hop in bed with me willingly."

36

1. In the spaces below, deal with three instances of your own behavior. Stick to describing what you do or fail to do rather than experiences or feelings. Choose situations and behaviors that are relevant to your role as helper.

a. **Vague.** _____

_____

b. **Concrete.** _____

_____

_____

_____

_____

a. **Vague.** _____

_____

b. **Concrete.** _____

_____

_____

_____

_____

a. **Vague.** _____

_____

b. **Concrete.** _____

_____

_____

_____

_____

2. Share one or two of these with your learning partner. Get feedback on how clear your second statement is. If you do not think that someone's statement is as clear as it might be, use probes to help your fellow trainee make his or her statement clearer.

EXERCISE 19: SELF-PROBING — SPEAKING CONCRETELY ABOUT YOUR FEELINGS

Feelings and emotions arise from experiences and behaviors. Therefore, it is unrealistic to talk about feelings without relating them to experiences or behaviors. However, in this exercise try to emphasize the feelings. Read the following examples.

Example 1: Jamie talking about how the training group affects him.

- **Vague statement of feelings.** "I get bothered in training groups."
- **Concrete statement of the same feelings.** "I feel hesitant and embarrassed whenever I want to give feedback to other trainees, especially if it is in any way negative. When the time comes, my heart beats faster and my palms sweat. I feel like everyone is staring at me."

Example 2: Renata talking about her relationship with her mother.

- **Vague statement of feelings.** "I feel unsettled at times with my mother."
- **Concrete statement of the same feelings.** "I feel guilty and depressed whenever my mother calls and implies that she's lonely. I then get angry with myself for giving in to guilt so easily. Then the whole day has a pall over it. I get nervous and irritable and show it to others."

1. In the spaces below, deal with three instances of your own feelings. Try to focus on feelings that you have some trouble managing and that could interfere with your role as helper.

a. **Vague.** _____

_____

b. **Concrete.** _____

_____

_____

_____

_____

a. **Vague.** _____

_____

b. **Concrete.** _____

_____

_____

_____

_____

a. **Vague.** _____

_____

b. **Concrete.** _____

_____

_____

_____

2. Share one or two of these with your learning partner. Get feedback on how clear your second statement is. If you do not think that someone's statement is as clear as it might be, use probes to help your fellow trainee make his or her statement clearer.

EXERCISE 20: SELF-PROBING — TELLING YOUR STORY CONCRETELY

In this exercise, you are asked to bring together all three elements — specific experiences, specific behaviors, and specific feelings — in talking about some personal concerns. Study the following examples.

Example:

- **Vague statement.** "Sometimes I'm a rather overly sensitive and spiteful person."
- **Concrete statement.** "I do not take criticism well. When I receive almost any kind of negative feedback, I usually smile and seem to shrug it off, but inside I begin to pout. Also, deep inside, I put the person who gave me the feedback on a 'list.' I say to myself that that person is going to pay for what he or she did. I find this hard to admit, even to myself. It sounds so petty. For instance, two weeks ago in the training group I received some negative feedback from Cindy. I felt angry and hurt because I thought she was my 'friend.' Since then I've tried to see what mistakes she makes here. I've been looking for an opportunity to get back at her. I've even felt bad because I haven't been able to catch her in any kind of glaring mistake. It goes without saying that I'm embarrassed to say all this."

Pick out the experiences, behaviors, and feelings in this example. Discuss to what degree the detail offered gets at the core of the problem situation.

1. Talk about two situations in terms of your own specific experiences, behaviors, and feelings. Deal with themes that relate to your potential effectiveness as a helper. Choose detail that gets at the core of the problem or unused opportunity.

a. **Vague.** _____

_____

b. **Concrete.** _____

_____

_____

_____

_____

_____

_____

a. **Vague.** _____

_____

b. **Concrete.** _____

_____

_____

_____

_____

_____

_____

_____

EXERCISE 21: PROBING FOR CLARITY OF EXPERIENCES, BEHAVIORS, AND FEELINGS

A probe is a statement or a question that invites a client to discuss an issue more fully. In the previous exercises you were asked to probe yourself. Probes are ways of getting at important details that clients do not think of or are reluctant to talk about. They can be used at any point in the helping process to clarify issues, search for missing data, expand perspectives, and point toward possible client actions. An overuse of probes can lead to gathering a great deal of irrelevant information. The purpose of a probe is not information for its own sake but data — experiences, behaviors, and feelings — that serve the process of problem management and opportunity development.

In this exercise, brief problem situations will be presented. Your job is to formulate two possible probes.

1. First respond with empathy.
2. Formulate a number of probes you might use with the client.
3. Jot down your reasons for using each probe.
4. In debriefing this exercise, give each other feedback on the quality of the empathy used.
5. Share with your learning partner your probes and your reasons for using each.
6. See if you can come to some agreement on which probes might work best.

Example: A man, 24, complains that he is severely tempted to go on experimenting sexually with women other than his wife: "Although I have not had an extended affair, I have had a few sexual encounters and feel that some day I will pursue a longer relationship. I don't blame myself or her. It's just that my feelings are so strong that she cannot satisfy me. I want more affection — really, more lust — than she can possibly give. I fear that no one woman will do so. I'm not sure what to think. I don't know whether I'm just being selfish or whether I just need to experiment more with relationships. I keep asking myself what all this means."

a. **Empathic statement.** "Your sexual urges are so strong that right now they're in the driver's seat. But you're not quite sure what all this is about and where it is leading."
b. **Possible probes.** "When you ask yourself what all this means, what are some of the answers you come up with?" "Tell me a bit about what you mean by 'selfish' here." "I'm not sure how much of this, if any, you've shared with your wife."

1. Grace, 19, an unmarried, first-year college student, comes to counseling because of an unexpected and unwanted pregnancy: "Right now I realize that the father could be either of two guys. That probably says something about me right there. I'm not sure what I want to do about the baby. I haven't told my parents yet, but I think that they will be sympathetic. But I've gotten myself into this mess, and I have to get myself out."

a. First, give an empathic response.

_____

_____

41

b. What are some probes that might help the client tell her story more fully?

_____

_____

_____

_____

2. You are a counselor in a halfway house. You are dealing with Tom, 44, who has just been released from prison where he served two years for armed robbery. He has been living at the halfway house for two weeks. That is the only offense for which he has ever been convicted. The halfway house experience is designed to help him reintegrate himself into society. Living in the house is voluntary. The immediate problem is that Tom came in drunk a couple of nights ago. He was supposed to be out on a job-search day. Drinking is against the rules of the house. When you talk to him, he mumbles vaguely, something about "still being confused."

a. First, give an empathic response.

_____

_____

_____

_____

b. What are some probes that might help the client tell his story more fully?

_____

_____

_____

_____

3. Arnie is a born-again Christian. He has begun to do a fair amount of preaching at his place of employment. While some of his co-workers sympathize with his views, others are turned off. Since he feels that he is being driven by a "clear vision," he becomes more and more militant. His supervisor has cautioned him a couple of times, but this has done little to change Arnie's behavior. Finally, he is given an ultimatum to talk to one of the counselors in the Employee Assistance Program about these issues or be suspended from his job. He says to the counselor, "I have a duty to spread the word. And if I have a duty to do so, then I also have the right. I'm a good worker. In fact, I believe in hard work. So it's not like I'm taking time off for the Lord's work. Now what's wrong with that?"

a. First, give an empathic response.

_____

_____

b. What are some probes that might help the client tell his story more fully?

_____

_____

_____

_____

Since, like empathy, probes can be used to good effect throughout the helping process, some work on probes will be included in later exercises.

EXERCISE 22: COMBINING EMPATHY WITH PROBES FOR CLARITY AND CLIENT ACTION

This exercise asks you to combine several skills — the ability to be empathic, to identify areas needing clarification, and to use probes to make clients aware of the need for action. Remember that each step of the helping process should be, all things considered, some kind of stimulus for client action, the "little" actions, as it were, that precede formal action based on a plan.

1. First reply to the client with empathy.
2. Identify an area needing exploration and clarification.
3. Use a probe to help the client explore or clarify some issue.
4. On the assumption that you have spent time understanding the client and helping him or her explore the problem situation through empathy and probes, indicate what action possibilities you might probe for.

Example 1: A law student, 25, is talking to a school counselor: "I learned yesterday that I've flunked out of school and that there's no recourse. I've seen everybody, but the door is shut tight. What a mess! I know I haven't gotten down to business the way I should. This is my first year in a large city, and there are so many distractions. And school is so competitive. I have no idea how I'll face my parents. They've paid for my college education and this year of law school. And now I'll have to tell them that it's all down the drain."

a. **Empathy.** "The whole situation sounds pretty desperate both here and at home. And it sounds so final."
b. **An area for probing.** It is not clear whom the client saw and what specific responses he received.

c. **Probe for clarity.** "I'm not sure who you mean by 'everybody,' and what doors were actually shut?"

d. **Probes for action possibilities.** Could he make further appeals? What advice would he give his brother if he were in this mess? How can he cut his losses? What might an honest appeal to his parents look like?

1. A high school senior to a school counselor: "My dad told me the other night that I looked relaxed. Well, that's a joke. I don't feel relaxed. There's a lull right now, because of semester break, but next semester I'm signed up for two math courses, and math really rips me up. But I need it for science, since I want to go into pre-med."

a. Empathy.

_____

_____

b. Fruitful area for probing.

_____

_____

c. Probe.

_____

_____

d. Probes for action possibilities.

_____

_____

_____

2. A woman, 27, talking to a counselor about a relationship that has just ended (she speaks in a rather matter-of-fact voice): "About three weeks ago I came back from visiting my parents who live in Nevada and found a letter from my friend Gary. He said that he still loves me but that I'm just not the person for him. In the letter he thanked me for all the good times we had together these last three years. He asked me not to try to contact him because this would only make it more difficult for both of us. End of story. I guess I've let my world collapse. People at work have begun complaining about me. And I've been like a zombie most of the time."

a. Empathy.

_____

_____

b. Fruitful area for probing.

_____

_____

c. Probe.

_____

_____

d. Probes for action possibilities.

_____

_____

_____

3. A married man, 25, talking to a counselor about trouble with his mother-in-law: "The way I see it, she is really trying to destroy our marriage. She's so conniving. And she's very clever. It's hard to catch her in what she's doing. You know, it's rather subtle. Well, I've had it! If she's trying to destroy our marriage, she's getting pretty close to achieving her goal."

a. Empathy.

_____

_____

b. Fruitful area for probing.

_____

_____

c. Probe.

_____

_____

d. Probes for action possibilities.

_____

_____

_____

4. A woman, 31, talking to an older woman friend: "I just can't stand my job any more! My boss is so unreasonable. He makes all sorts of silly demands on me. The other women in the office are so stuffy, you can't even talk to them. The men are either very blah or after you all the time, you know, on the make. The pay is good, but I don't think it makes up for all the rest. It's been going on like this for almost two years."

a. Empathy.

_____

_____

b. Fruitful area for probing.

_____

_____

c. Probe.

_____

_____

d. Probes for action possibilities.

_____

_____

_____

_____

5. A man, 45, who has lost his wife and home in a tornado, has been talking about his loss to a social worker: "This happened to a friend of mine in Kansas about ten years ago. He never recovered from it. His life just disintegrated, and nobody could do anything about it. . . . It was like the end of the world for him. You never think it's going to happen to you. I know I belong here. But this kind of thing makes me think I don't."

a. Empathy.

_____

_____

b. Fruitful area for probing.

_____

_____

c. Probe.

_____

_____

d. Probes for action possibilities.

_____

_____

_____

Supplemental exercises are found in Appendix Two.

PART THREE

STAGE I OF THE HELPING MODEL AND ADVANCED COMMUNICATION SKILLS

In Parts Three and Four of this manual the exercises focus on the stages and steps of the helping process together with the advanced communication skills needed to help clients get value from each session and make progress. In Part Three the focus is on the three steps of Stage I and steps of the helping model. These exercises will enable you to help your clients:

- identify and clarify problem situations and missed opportunities.
- challenge themselves to develop new and more useful perspectives on themselves and their problems.
- work on key issues, that is, issues that will make a difference in their lives.
- commit themselves to work on these issues in pragmatic, action-oriented ways.

As indicated in the text, the steps or tasks of Stage I are not only interrelated but apply to all stages and steps of the helping process. That is, throughout the counseling process, clients need to be helped to tell their stories, challenge themselves, and work on things that make a difference.

Section 7
STEP I-A: HELPING CLIENTS TELL THEIR STORIES

The communication skills reviewed in Part Two are tools you need to deliver the stages and steps of the helping model. Stage I deals with problem identification and clarification. Step A in this stage deals with helping clients tell their stories, that is, discuss the problem situations, anxieties, and concerns that bring them (or get them sent) to the helper in the first place. Read Chapter Seven in *The Skilled Helper* before doing the exercise in this section.

EXERCISE 23: HELPING ONE ANOTHER TELL YOUR STORIES

In the exercises you have done to this point you have already been helping clients tell their stories through the use of empathy and probes. It is time now to switch from writing to dialogue.

1. Form groups of three.
2. In each group there are three roles: client, helper, and observer.
3. Before the dialogue between client and helper begins, the trainee in the client role gives a very brief summary of the problem situation or unused opportunity he or she is going to explore.

Example: Enrico, the trainee playing the role of client, gives the following summary: "I'm going to discuss a problem I'm having with my parents. I'm an only child, and they simply won't let go. Don't get me wrong. They're very nice about it. But I'm 20. We're not a wealthy family. So, for the time being, I have to live at home. Also I believe that becoming my own person is directly related to becoming an effective helper."

4. The helper begins by saying: "Against that background, Enrico, what would you like to focus on today?"
5. Then client and helper engage in a dialogue. The helper, using whatever combination of empathy and probes he or she sees necessary, helps the client tell his or her story clearly and in useful detail.
6. Spend about six minutes in the dialogue. The observer is timekeeper.
7. After about six minutes the observer stops the dialogue.
8. The client gives the helper feedback on how effectively he or she has aided the story telling process.
9. Then the observer complements what the client says with his or her feedback.
10. The process continues until each member has had the opportunity to play each role.
11. Then proceed to a second round. In the second round the trainee in the client role is asked to summarize the first dialogue and then move on from there.

Section 8
STEP I-B: HELPING CLIENTS CHALLENGE THEMSELVES

Chapter Eight in *The Skilled Helper* discusses the role of challenging in the helping process. Helping clients challenge themselves both to participate actively in the helping process and to act on what they learn adds great value. While challenge without understanding is abrasive, understanding without challenge is anemic. Before doing the following exercise, read Chapter Eight.

EXERCISE 24: IDENTIFYING DIFFERENT AREAS NEEDING CHALLENGE

There are various ways in which clients need to be helped to challenge themselves in order to develop the kinds of new perspectives that lead to constructive problem-managing and opportunity-developing action. In this exercise you are asked to do two things.

1. Read the case and determine in what way or in what area this client might benefit from challenge.
2. Sketch out briefly what some helpful new perspectives might look like.
3. Share your observations with a learning partner. Note the similarities and the differences among your suggestions.
4. Discuss how you might proceed in each case, finding ways to help these clients challenge themselves.

Example: Hannah, a new counselor working in a hospital setting for a number of months, has begun to confide in one of her colleagues who has worked in health-care facilities for about seven years. Hannah has intimated, however tentatively, that she feels that some of the medical doctors tend to dismiss her because she is new, because she is a counselor and not a medical practitioner, and because she is a woman. One doctor has been particularly nasty, even though he masks his abusiveness under his brand of "humor." One day her colleague says to her, "Tell 'em what you think!" Hannah replies: "That will only make them worse. I think it more important to stand up for my profession by just being who I am and helping patients as much as possible. Their meanness is a punishment in itself. Anyway, it's a game that doctors play with everyone else in the hospital. That kind of thing has been going on forever. It's not such a big deal anyway."

- **What might be challenged here?** Under the guise of "reasonableness," Hannah seems to be into excuse making. She might be helped to explore her reasons for not doing anything about the situation. There could be reasons under her reasons that might reveal other issues in the problem situation.
- **What would some helpful new perspectives look like?** Hannah might be helped to see that accepting the status quo goes counter to the kind of professionalism that should characterize the hospital and that she prides herself in. She may also be helped to see that the hospital's caste system might interfere with team approaches to helping patients.
- **What actions might such a perspective lead to?** Hannah might bring the core issues before the hospital senate, she might quietly lobby some of her colleagues to see what kind of consensus there is, she might talk to the abusive doctor privately and let him know how she feels, she might confide in one of the doctors who seems to be free of these prejudices in order to see what course of action she might follow, and so forth.

It goes without saying that the best way of doing this is through actual interaction with these clients. At this stage, however, you are being asked to do some educated guessing.

1. Dan is in a coaching/counseling session with his supervisor, June. A project is behind schedule and a key customer is complaining. Dan was chosen to be the project leader because he had manifested "leadership" qualities in his work. He says something like this: "Well, this is one of the first team-focused projects we've had. We're not used to working like this. The key

people in design and engineering have been slow to come to the table. They keep telling me that they can't let other projects slip. I know the customer wants the prototype by the end of the month, but that's his schedule, not necessarily reality. He's so pushy. Lately, it's hard even taking his calls."

a. What might be challenged here?

_____

_____

_____

b. What would some helpful new perspectives look like?

_____

_____

_____

c. What actions might such a perspective lead to?

_____

_____

_____

2. Claudette and Jim have tried unsuccessfully to have children. Doctors have examined both of them and now offer little hope even with new medical techniques. They have both been seeing a counselor, but in this session Claudette, who is much less reconciled to their fate than Jim is, is seeing the counselor alone. She says, "It's just not fair. Doctors talk about infertility as a clinical phenomenon, not a human reality. Jim and I love one another. We want to have children. We want the sex we have together to lead to life. We'd make very good parents. We've done everything the doctors have told us to do, and still nothing."

a. What might be challenged here?

_____

_____

_____

b. What would a helpful new perspective look like?

_____

c. What actions might such a perspective lead to?

_____

_____

_____

3. Len, married with three teenage children, lost all the family's savings in a gambling spree. Under a lot of pressure from both his family and his boss, he started attending Gamblers Anonymous meetings. He seemed to recover; that is, he stopped betting on horses and ball games. A couple of years went by, and Len stopped going to the GA meetings because he "no longer needed them." He even got a better job and was recovering financially quite well. But his wife began to notice that he was on the phone a great deal with his broker. She confronted him, and he reluctantly agreed to a session with one of the friends he made at GA. He says, "She's worried that I'm gambling again. You know, it's really just the opposite. When I was gambling, I was financially irresponsible. I lost our future on horses and ball games. But now I'm taking a very active part in creating our financial future. Every financial planner will tell you that investing in the market is central to sound financial planning. I'm a doer. I've taken a very active role."

a. What might be challenged here?

_____

_____

_____

b. What would a helpful new perspective look like?

_____

_____

_____

c. What actions might such a perspective lead to?

_____

_____

_____

_____

4. Stepan, a refugee who had been brutalized by a political regime in his native country, has been seeing a counselor in a center that specializes in helping victims of torture. He has been in the country for two years. He has a decent job. Even though he is now secure, at least in some sense of that term, he has found it very difficult to establish relationships with other refugees, with members of the immigrant community, or with native-born Americans. He intimates that he is quite lonely, but dismisses loneliness as "inconsequential" when the counselor brings it up. In one session, he says: "Let's not talk about my so-called loneliness when there is a great world loneliness out there. The loneliness that brutality creates; now that's something that counts. I fill my days quite well. When it is time, I'll seek out others, but I need people who can see the world as I see it. As it really is."

a. What might be challenged here?

_____

_____

_____

b. What would a helpful new perspective look like?

_____

_____

_____

c. What actions might such a perspective lead to?

_____

_____

_____

Section 9
THE SKILLS OF EFFECTIVE CHALLENGING

Challenging skills include: (1) information sharing, (2) advanced empathy, (3) helper self-disclosure, (4) immediacy, and (5) summarizing. The purpose of these skills is to help clients get in touch with blind spots and develop the kind of new perspectives or behavioral insights needed to complete the clarification of a problem situation and to move on to developing new scenarios, setting problem-managing goals, developing strategies, and moving to action. Challenging skills and processes are useful to the degree that they serve the stages and steps of the helping process.

1. INFORMATION SHARING

As noted in the text, sometimes clients do not get a clear picture of a problem situation or manage it effectively because they lack information needed for clarity and action. Information can provide clients with some of the new perspectives they need to see problem situations as manageable. Giving clients problem-clarifying information or helping them find it themselves is not, of course, the same as advice giving or preaching.

EXERCISE 25: INFORMATION FOR MANAGING YOUR OWN PROBLEMS/OPPORTUNITIES

In this exercise you are asked to review the problem areas you have chosen to deal with during this training program. Choose two areas and ask yourself whether there is some kind of information that would help you understand your problem more thoroughly or help you do something to manage it more effectively.

Example: Jess, a counselor trainee, 30, has been married a little over a year. He has just quit work to start full time in school in a counselor training program. He is having trouble with his marriage. He is having second thoughts about quitting his job and entering the program. He has had no background in psychology previous to this. In college he majored in history. He asks himself: "What kind of information would help me see my concerns more clearly and manage them more effectively?"

- I need information about what job opportunities there are for people like myself with an M.A. in counseling psychology. This will help me to commit myself to the program more fully.
- It would be helpful for me (and my wife) to know more about what kinds of pitfalls exist for a couple in the first two years of marriage. I have a feeling that some of the problems we are having are not uncommon for "beginners."
- I don't know what the ordinary developmental challenges are for someone my age. My wife is 24. We are at different life-cycle stages. She seems to be going through stuff I've already seen. I feel uprooted. Some applied developmental psychology would help me.

"Now that I write this down, it's clear that I can get a lot of this information for myself as part of this counseling psychology program."

Jess found the answers to some of his questions quite challenging. For instance, it was disturbing to learn that many people in the helping professions considered the M.A. a second-class degree, while for him it would be quite an achievement. He knew that he would have to talk to M.A. graduates working in helping settings to see this issue through their eyes. He also knew that he had to find out what the future held for helpers with an M.A. He also discovered that for many couples the first two years of marriage can be the trickiest.

1. Choose two issues, opportunities, or concerns you are working on in your own life and, like Jess, ask yourself what kind of information would help you see yourself and your problem situation more clearly and do something about managing it.

2. Share what you have learned with a learning partner and point out the ways in which you have been challenged by the information you have dug up.

a. Problem or opportunity area # 1.

_____

_____

b: Information needed.

_____

_____

c. Critical findings.

_____

_____

_____

d. Ways in which information challenged me.

_____

_____

_____

_____

a. Problem or opportunity area # 2.

_____

_____

_____

b. Information needed.

_____

_____

c. Critical findings.

_____

_____

_____

d. Ways in which information challenged me.

_____

_____

_____

_____

3. Finally, debrief your findings with your learning partner. Discuss the power of relevant information.

EXERCISE 26: INFORMATION LEADING TO NEW PERSPECTIVES AND ACTION

In this exercise you are asked to consider what kind of information you might give or help clients obtain that would help them see the problem situations they are facing more clearly. Information can, somewhat artificially, be divided into two kinds: (a) information that helps clients understand their difficulties better, and (b) information about what actions they might take.

Example: Tim was a bright, personable young man. During college he was hospitalized after taking a drug overdose during a bout of depression. He spent six months as an in-patient. He was assigned to "milieu therapy," an amorphous mixture of work and recreation designed more to keep patients busy than to help them grapple with their problems and engage in constructive change. He was given drugs for his depression, seen occasionally by a psychiatrist, and assigned to a therapy group that proved to be quite aimless. After leaving the hospital, his confidence shattered, he left college and got involved with a variety of low-paying, part-time jobs. He finally finished college by going to night school, but he avoided full-time jobs for fear of being asked about his past. Buried inside him was the thought, "I have this terrible secret that I have to keep from everyone." A friend talked him into taking a college-sponsored communication-skills course one summer. The psychologist running the program, noting Tim's rather substantial natural talents together with his self-effacing ways, remarked to him one day, "I wonder what kind of ambitions you have." In an instant Tim realized that he had buried all thoughts of ambition. After all, he didn't "deserve" to be ambitious. Tim, instinctively trusting him, divulged his "terrible secret" for the first time.

Tim and the psychologist had a number of meetings over the next few years. Challenge became an important part of the helping process. In some of the early sessions, Rick, Tim's counselor, provided him with information that proved challenging in a number of areas. For instance, when it came to employment, the counselor helped Tim find out that:

- he was probably more intelligent and talented than he realized and that he was underemployed,
- employers do not necessarily dig deeply into the background of prospective employees,
- privacy laws protect often protect prospective employees from damaging disclosures,
- enlightened employers look benignly on the peccadillos and developmental crises of a prospective employee's adolescent years, and
- the job market was strong and growing stronger so that people with Tim's skills and potential were in high demand.

All this challenged Tim to look at himself in a different light, that is, as a potential winner instead of a loser, and to move more aggressively into job hunting.

In the following cases you are not asked to be an expert. Rather, from a lay person's or a common-sense point of view, you are asked to indicate what kinds of information you believe might help the client understand and manage the problem situation more clearly. Specific, expert information is not essential. What blind spots do you think the client has? How do you think information might help? In what ways would the information be challenging?

1. Indicate what information could help the client get some new perspectives and move to action.
2. Discuss with your learning partner the kinds of new perspectives you would be trying to help the client develop and the actions the client might take based on the new information.

1. A man, 26, has just been sentenced to five years in a penitentiary. He is talking to a chaplain-counselor who has worked for the past ten years at the penitentiary to which the man has been assigned. The chaplain also counseled the man during his trial. The man is deathly afraid of going to prison and has even talked about taking his own life.

What information might help this client develop new perspectives? What actions might these new perspectives lead to?

_____

_____

_____

_____

2. A woman, 45, has learned that she has cancer and will soon undergo a mastectomy. She has been put in touch with a self-help group composed of women who have had this operation. She is now talking to one of the members of this group, but she hasn't made up her mind to attend.

What information might help this client develop new perspectives? What actions might these new perspectives lead to?

_____

_____

_____

_____

3. A woman, 28, has been raped and is talking to a counselor at a counseling center for victims of rape. The rape took place two days ago. She has not yet reported the rape to the police. She says that she wants to make as little fuss as possible.

What information might help this client develop new perspectives? What actions might these new perspectives lead to?

_____

_____

_____

_____

4. A man, 34, comes for counseling because he fears his drinking might be getting out of hand. He's been drinking heavily for several years and recently has had some physical symptoms that he hasn't experienced before, for instance, blackouts. He has managed to remain a secret drinker. At his job he works by himself and has not had more "sick" days than the average. None of his immediate family or friends has been an alcoholic.

What information might help this client develop new perspectives? What actions might these new perspectives lead to?

_____

_____

_____

_____

_____

5. A man, 41, comes to counseling because he fears he is going crazy. He has a number of problems. His marriage has deteriorated in the past year or so. He and his wife relate poorly to each other. He sees his teenage son and daughter drifting away from him, and he doesn't understand this. He fears that his son might be taking drugs on occasion but has not confronted him. He is drinking more than he should. He goes in and out of bouts of mild depression. He has begun to steal little things from stores, not because he needs them but because "it gives him a lift."

What information might help this client develop new perspectives? What actions might these new perspectives lead to?

_____

_____

_____

_____

_____

Supplemental exercises can be found in Appendix Two.

2. ADVANCED EMPATHY

Advanced empathy, described most simply, means sharing **hunches** about clients and their overt and covert experiences, behaviors, and feelings that you feel will help them see their problems and concerns more clearly and help them move on to developing new scenarios, setting goals, and acting. Such hunches, of course, must be based on the helper's interactions with the client in which active listening, understanding, empathy, and probing play a large part. Advanced empathy as hunch sharing can be expressed in a number of ways. Some of these are reviewed briefly below. Before doing the following exercises, however, review the section of advanced empathy in the text.

Advanced Empathy as Educated Hunches

- Hunches that help clients see the **bigger picture**. In this example, the counselor is talking to a client who is at odds with his wife's brother: "The problem doesn't seem to be just your attitude toward your brother-in-law any more; your resentment seems to be spreading to his friends. How do you see it?"
- Hunches that help the clients see what they are expressing **indirectly** or merely **implying**. Example: The counselor is talking to a client who feels that a friend has let her down: "I think I might also be hearing you say that you are more than disappointed — perhaps a bit hurt and angry."
- Hunches that help clients draw **logical conclusions** from what they are saying. A manager is counseling one of his team members: "From all that you've said about her, it seems that you

are also saying that right now you resent having to work with her at all. I know you haven't said that directly. But I'm wondering if you are feeling that way about her."

- Hunches that help clients open up areas they are only **hinting** at. In this case a school counselor is talking to a senior in high school: "You've brought up sexual matters a number of times, but you haven't pursued them. My guess is that sex is a pretty important area for you but perhaps pretty touchy, too."
- Hunches that help clients see things they may be **overlooking**. A counselor is talking to a minister: "I wonder if it's possible that some people take your wit too personally, that they see it, perhaps, as sarcasm rather than humor."
- Hunches that help clients identify **themes**. In this case a counselor is talking to a woman who has been abused by her husband: "If I'm not mistaken, you've mentioned in two or three different ways that it is sometimes difficult for you to stick up for your own legitimate rights. For instance . . . ."
- Hunches that help clients **own** only partially expressed experiences, behaviors, and/or feelings. Example: "You sound as if you have already decided to quit. Tell me if I'm going too far."

Hunches should be based on your experience of your clients — their experiences, behaviors, and emotions both within the helping sessions themselves and in their day-to-day lives. Do not base your hunches on "deep" psychological theories. Later on, you will be asked to identify the experiential and behavioral clues on which your hunches are based.

EXERCISE 27: ADVANCED ACCURATE EMPATHY — HUNCHES ABOUT YOURSELF

One way to get an experiential feeling for advanced empathy is to explore *at two levels* some situation or issue in your own life that you would like to understand more clearly. One level of understanding could be called the surface level. The second could be called a more objective or a deeper level.

1. Read the examples given below.

Example 1: A man, 25, in a counselor training program believes that his experience in the training group is giving him some second thoughts about his ability and willingness to get close to others.

- **Level-1 understanding:** "I like people, and I show this by my willingness to work hard with them. For instance, in this group I see myself as a hard worker. I listen to others carefully, and I try to respond carefully. I see myself as a very active member of this group. I take the initiative in contacting others. I like working with the people here."
- **Below-the-surface understanding:** "If I look closer at what I'm doing here, I realize that underneath my 'hardworking' and competent exterior I am uncomfortable. I come to these sessions with more misgivings than I have admitted, even to myself. My hunch is that I have some fears about human closeness. I am afraid, both here and in a couple of relationships outside the group, that someone is going to ask me for more than I want to give. This keeps me on edge here. It keeps me on edge in a couple of relationships outside."

Example 2: A woman, 33, in a counselor training group, sees that her experience in the group is making her explore her attitude toward herself. It might not be as positive as she thought.

She sees this as something that could interfere with her effectiveness as a counselor.

- **Level-1 understanding:** "I like myself. I base this on the fact that I seem to relate freely to others. There are a number of things I like specifically about myself. I'm fairly bright. And I think I can use my intelligence to work with others as a helper. I work hard. I'm demanding of myself, but I don't place unreasonable demands on others."
- **Below-the-surface understanding:** "If I look more closely at myself, I see that when I work hard it is because I feel I have to. My hunch is that 'I have to' counts more in my hard work than 'I want to.' I get pleasure out of working hard, but it also keeps me from feeling guilty. If I don't work 'hard enough,' then I can feel guilty or down on myself. I am beginning to feel that there is too much of the 'I must be a perfect person' in me. I judge myself and others more harshly than I care to think."

2. Choose some issue, topic, situation, or relationship that you have been investigating and which you would like to understand more fully with a view to taking some kind of action on it. As usual, choose issues that you are willing to share with the members of your training group, and try to choose issues that might affect the quality of your counseling.
3. Briefly describe the issue, as in the examples.
4. Then give a "surface-level" description of the issue.
5. Next, share some hunch you have about yourself that relates to that issue, a hunch that "goes below the surface," as it were. This will help you get in touch with possible blind spots. Try to develop a new perspective on yourself and that issue, one that might help you see the issue more clearly so that you might begin to think of how you might act on it.

Area # 1.

_____

_____

_____

Area # 2.

_____

_____

_____

6. Share your insights with your learning partner.

EXERCISE 28: THE DISTINCTION BETWEEN BASIC AND ADVANCED EMPATHY

In this exercise, assume that the helper and the client have established a good working relationship, that the client's concerns have been explored from his or her perspective, and that the client needs to be challenged to see the problem situation from some new perspective or to act on the insights that have been developed.

1. In each instance, imagine the client speaking directly to you.
2. In (a) respond to what the client has just said with basic empathy. Use the formula or your own words.
3. Next, formulate one or two hunches about this person's experiences, behaviors, or feelings, hunches that, when shared, would help promote understanding and/or action. Use the material in the "context" section together with the client's words to formulate your hunches. Ask yourself: "On what clues am I basing this hunch?"
4. Then in (b) respond with some form of advanced empathy; that is, share some hunch that you believe will be useful for him or her. Share it in a way that will not put the client off.

Example: A man, 48, husband and father, is exploring the poor relationships he has with his wife and children. In general, he feels that he is the victim, that his family is not treating him right (that is, like many clients, he emphasizes his experience rather than his behavior). He has not yet examined the implications of the ways he behaves toward his family. At this point he is talking about his sense of humor. He says: "For instance, I get a lot of encouragement for being witty at parties. Almost everyone laughs. I think I provide a lot of entertainment, and others like it. It also works on my job. But this is another way I seem to flop at home. When I try to be funny, my wife and kids don't laugh, at least not much. At times they even take my humor wrong and get angry. I actually have to watch my step in my own home."

a. **Basic empathy.** "It's irritating when your own family doesn't seem to appreciate what you see as one of your real talents, one that others do appreciate."
b. **Hunch.** The family wants a husband and father, not a humorist. His humor, especially at home, is not as harmless as he thinks.
c. **Advanced empathy.** "I wonder whether their reaction to you could be interpreted differently. For instance, they might not want an entertainer at home, but just a husband and father. You know, just you."

1. A first-year engineering graduate student has been exploring his disappointment with himself and with his performance in school. He has explored such issues as his dislike for the school and for some of the teachers with his counselor. He says: "I just don't have much enthusiasm. My grades are just okay, maybe even a little below par. I know I could do better if I wanted to. I don't know why my disappointment with the school and some of the faculty members can get to me so much. It's not like me. Ever since I can remember — even in primary school, when I didn't have any idea what an engineer was — I've wanted to be an engineer. Theoretically, I should be as happy as a lark because I'm in graduate school, but I'm not."

a. Basic empathy.

_____

_____

b. Your hunch and your reason for it.

_____

_____

c. Advanced empathy.

_____

_____

_____

2. This man, now 64, retired early from work when he was 62. He and his wife wanted to take full advantage of the years they had left. But his wife died a year after he retired. At the urging of friends he has finally come to a counselor. He has been exploring some of the problems his retirement has created for him. His two married sons live with their families in other cities. In the counseling sessions he has been dealing somewhat repetitiously with the theme of loss. He says: "I seldom see the kids. I enjoy them and their families a lot when they do come. I get along real well with their wives. But now that my wife is gone . . . (pause) . . . and since I've stopped working . . . (pause) . . . I seem to just ramble around the house aimlessly, which is not like me at all. I suppose I should get rid of the house, but it's filled with a lot of memories—bittersweet memories now. There were a lot of good years here. The years seem to have slipped by and caught me unawares."

a. Basic empathy.

_____

_____

b. Your hunch and your reason for it.

_____

_____

_____

c. Advanced empathy.

_____

_____

_____

3. A single woman, 33, is talking to a minister about the quality of her social life. She has a very close friend and she counts on her a great deal. She is exploring the ups and downs of this relationship. In the counseling sessions this woman comes on a bit loud and somewhat aggressive. She says: "Ruth and I are on again, off again with each other lately. When we're on,

it's great. We have lunch together, go shopping, all that kind of stuff. But sometimes she seems to click off. You know, she tries to avoid me. But that's not easy to do. I keep after her. She's been pretty elusive for about two weeks now. I don't know why she runs away like this. I know we have our differences. She quieter, and I'm the louder type. But our differences don't ordinarily seem to get in the way."

a. Basic empathy.

_____

_____

b. Your hunch and your reason for it.

_____

_____

_____

c. Advanced empathy.

_____

_____

_____

4. A man, 40, is talking to a marriage counselor. This is the third time he has come to see the counselor over the past four years. His wife has never come with him. The other times he spent only a session or two with the counselor and then dropped out. In this session he has been talking a great deal about his latest annoyances with his wife. He says: "I could go on telling you what she does and doesn't do. It's a litany. She really knows how to punish, not only me but others. I don't even know why I keep putting up with it. I want her to come to counseling, but she won't come. So, here I am again, in her place."

a. Basic empathy.

_____

_____

b. Your hunch and your reason for it.

_____

_____

c. Advanced empathy.

_____

_____

_____

5. A nun, 44, a member of a counselor training group, has been talking about her dissatisfaction with her present job. Although a nurse, she is presently teaching in a primary school because, she says, of the "urgent needs" of that school. When pressed, she refers briefly to a history of job dissatisfaction. In the group she has shown herself to be an active, intelligent, and caring woman who tends to speak and act in self-effacing ways. She mentions how obedience has been stressed throughout her years in the religious order. She does mention, however, that things have been "letting up a bit" in recent years. The younger sisters don't seem to be as preoccupied with obedience as she is. She says: "The reason I'm talking about my job is that I don't want to become a counselor and then discover it's another job I'm dissatisfied with. It would be unfair to the people I'd be working with and unfair to my religious order, which is paying for my education. Of course, I have no iron-clad assurance that I'll be put in a job that will enable me to use my counselor training."

a. Basic empathy.

_____

_____

b. Your hunch and your reason for it.

_____

_____

_____

_____

c. Advanced empathy.

_____

_____

_____

Supplemental exercises are found in Appendix Two.

EXERCISE 29: CHALLENGING DISCREPANCIES IN YOUR OWN LIFE

Most of us face a variety of self-defeating discrepancies in our lives besides the discrepancies that involve unused strengths and resources. We all allow ourselves, to a greater or lesser extent, to become victims of our own prejudices, smoke screens, distortions, and self-deceptions. In this exercise you are asked to confront some of these, especially the kinds of discrepancies that might affect the quality of your helping or the quality of your membership in the training group.

Example 1

- **The issue.** "I am very controlling in my relationships with others."
- **The description.** "I am very controlling in my relationships with others. For instance, in social situations I manipulate people into doing what I want to do. I do this as subtly as possible. I find out what everyone wants to do, and then I use one against the other and gentle persuasion to steer people in the direction in which I want to go. In the training sessions I try to get people to talk about problems that are of interest to me. I even use empathy and probes to steer people in directions I might find interesting. All this is so much a part of my style that usually I don't even notice it. I see this as selfish, but I experience little guilt about it."

Example 2

- **The issue.** This trainee confronts her need for approval from others.
- **The description.** "Most people see me as a 'nice' person. Part of this I like, part of it is a smoke screen. Being nice is my best defense against harshness and criticism from others. I'm cooperative. I compliment others easily. I'm not cynical or sarcastic. I've gotten to enjoy this kind of being 'nice.' I find it rewarding. But it also means that I seldom talk about ideas that might offend others. My feedback to others in the group is almost always positive. I let others give feedback on mistakes. Outside the group I steer clear of controversial conversations. But I'm beginning to feel very bland."

Now confront yourself in two areas that, if dealt with, will help you be a more effective trainee and helper. In the description section be as specific as you can. Describe specific experiences, behaviors, feelings. Give brief examples.

a. Issue # 1.

_____

_____

b. Descriptive self-challenge.

_____

_____

_____

_____

_____

_____

_____

a. Issue # 2.

_____

_____

b. Descriptive self-challenge.

_____

_____

_____

_____

_____

_____

_____

_____

EXERCISE 30: CHALLENGING YOUR OWN STRENGTHS

One of the best forms of challenge is to invite clients to examine unused strengths and resources that could be used to manage some problem situation more effectively. In this exercise you are asked to confront yourself with respect to your own unused or underused strengths and resources. The discrepancy is that you have a resource but do not use it effectively.

Example:

·   **Problem situation.** "My social life is not nearly as full as I would like it to be."
·   **Description of unused strengths or resources.** "I have problem-solving skills, but I don't apply

67

them to the practical problems of everyday life such as my less than adequate social life. Instead of defining goals for myself (making acquaintances, developing friendships) and then seeing how many different ways I could go about achieving these goals, I wait around to see if something will happen to make my social life fuller. I remain passive even though I have the skills to become active."

1. Briefly identify some problem situations.
2. As in the example, indicate how some unused or underused strength or strengths you have could be applied to the management of each problem situation.
3. In addition, indicate some actions you might take to make use of underused strengths.

a. Problem situation # 1.

_____

_____

_____

b. Underused strengths as applied to the problem situation and actions to be taken.

_____

_____

_____

_____

_____

a. Problem situation # 2.

_____

_____

_____

b. Underused strengths as applied to the problem situation and actions to be taken.

_____

_____

_____

_____

_____

_____

_____

3. HELPER SELF-DISCLOSURE

Although helpers should be **ready** to make disclosures about themselves that would help their clients understand their problem situations more clearly, they should do so only if such disclosures do not upset their clients or distract their clients from the work they are doing. Read the text on helper self-disclosure before doing this exercise.

EXERCISE 31: EXPERIENCES OF MINE THAT MIGHT BE HELPFUL TO OTHERS

In this exercise you are asked to review some problems in living that you feel you have managed or are managing successfully. Indicate what you might share about yourself that would help a client with a similar problem situation understand that problem situation or some part of it more clearly and move on to problem-managing action. That is, what might you share of yourself that would help the client move forward in the problem-managing process?

Trainee Example 1: "In the past I have been an expert in feeling sorry for myself whenever I had to face any kind of difficulty. I know very well the rewards of seeing myself as victim. I used to fantasize myself as victim as a form of daydreaming or recreation. I think many clients get mired down in their problems because they let themselves feel sorry for themselves the way I did. I think I can spot this tendency in others. When I see this happening, I think I could share brief examples from my own experience and then ask clients to see if what I was doing squares with what they see themselves doing now."

Trainee Example 2: "I have been addicted to a number of things in my life and I see a common pattern in different kinds of addiction. For instance, I have been addicted to alcohol, to cigarettes, and to sleeping pills. I have also been addicted to people. By this I mean that at times in my life I have been a very dependent person, and I found the same kind of symptoms in dependency that I did in addiction. I know a lot about the fear of letting go and the pain of withdrawal. I think I could share some of this in ways that would not accuse or frighten clients or distract them from their own concerns."

1. List two areas in which you feel you have something to share that might help clients who have problems in living similar to your own. Just briefly indicate the area.

Area # 1.

_____

_____

Area # 2.

_____

_____

2. On separate paper make more extended comments in each area, comments similar to those in the examples.
3. Share one or two with your learning partner. Give and get feedback on the usefulness of the disclosures.

4. IMMEDIACY: EXPLORING RELATIONSHIPS

As noted in the text, your ability to deal directly with what is happening between you and your clients in the helping sessions themselves is an important skill. **Relationship** immediacy refers to your ability to review the history and present status of your relationship with other members of your group — who are both your fellow trainees and your "clients" — in concrete behavioral ways. **Here-and-now** immediacy refers to your ability to deal with a particular situation that is affecting the ways in which you and another person are relating right now, in this moment.

Immediacy is a complex skill. It involves: (1) revealing how you are being affected by the other person, (2) exploring your own behavior toward the other person, (3) sharing hunches about his or her behavior toward you or pointing out discrepancies, distortions, smoke screens, and the like, and (4) inviting the other person to explore the relationship with a view to developing a better working relationship. For instance, if you see that a client is manifesting hostility toward you in subtle, hard-to-get-at ways, you may: (1) let the client know how you are being affected by what is happening in the relationship (that is, you share your experience), (2) explore how you might be contributing to the difficulty, (3) describe the client's behavior and share reasonable hunches about what is happening (challenge), and (4) invite the client to examine in a direct way what is happening in the relationship. Immediacy involves collaborative problem solving with respect to the relationship itself.

EXERCISE 32: IMMEDIACY IN YOUR INTERPERSONAL LIFE

In this exercise you are asked to review some issues that remain "unfinished" between you and others outside the training group.

1. Think of people in your life with whom you have some unresolved or undealt-with "you-me" issues (relatives, friends, intimates, co-workers, and so forth).
2. Briefly indicate what the issue is.
3. Imagine yourself talking with one of these individuals face to face.
4. Be immediate with this person with a view to instituting the kind of dialogue that would help the two of you grapple with the issue that concerns you. Your immediacy statement should include (a) the issue and how it is affecting you, (b) some indication of how you might be contributing to the difficulty, (c) some kind of concrete challenge, and (d) an invitation to the other person to engage in dialogue with you on this issue.
5. Remember that initial challenges should be appropriately tentative.

Example 1

- **The issue.** A trainee sees herself speaking to a friend outside the group. She is dissatisfied with the depth of sharing in the relationship. She is hesitant about revealing her own deeper thoughts, values, and concerns.
- **The trainee talking directly to her friend.** "I'm a bit embarrassed about what I'm going to say. I think we enjoy being with each other. But I feel some reluctance in talking to you about some of my deeper thoughts and concerns. And, if I'm not mistaken, I see some of the same kind of reluctance in you. For instance, the other day both of us seemed to be pretty awkward when we talked a bit about religion. We dropped the subject pretty quickly. I'm embarrassed right now because I feel that I may be violating the 'not-too-deep' rule that we've perhaps stumbled into. I'm wondering what you might think about all this."

Example 2

- **The issue.** A trainee is speaking about her relationship with her boss. She feels that he respects her but, because she is a woman, he does not think of her as a prospect for managerial training.
- **The trainee, talking to her boss.** "I think you see me as a good worker. As far as I can tell, you and I work well together. Even though you're my boss, I see a sort of equality between us. I mean that you don't push your boss role. And yet something bothers me. Every now and then I pick up clues that you don't think of me when you're considering people for managerial training slots. You seem to be very satisfied with my work, but part of that seems to be being satisfied with keeping me in the slot I'm in. I don't see you as offensively sexist at all, but something tells me that you might unconsciously think of men for training slots before women. Maybe it's part of the culture here. It would be helpful for me if we could explore this a bit."

Now write out two statements of immediacy dealing with people in your life outside the training group. Choose people and issues that you would be willing to discuss in the group. Obviously, you need not reveal the identity of the people involved.

1. The issue.

Write out a face-to-face statement on separate paper.

2. **The issue.**

Write out a face-to-face statement on separate paper.

EXERCISE 33: RESPONDING TO SITUATIONS CALLING FOR IMMEDIACY

In this exercise a number of client-helper situations calling for some kind of immediacy on your part are described. You are asked to consider each situation and respond with some statement of immediacy. Consider the following example.

Example

- **Situation.** This client, a man of 44, occasionally makes snide remarks either about the helping profession itself or some of the things that you do in your interactions with him. At times he is cooperative, but at times he asks you to make decisions for him. "Just tell me the best way for me to tell my boss that he's an idiot." He makes remarks about you personally at times, " I bet you've got a lot of friends." Or, on being invited to challenge himself by you, "I hope you're a little more caring with your friends."
- **Immediacy response.** "Tom, let's stop a minute and explore what's happening between you and me in our sessions. . . . (Tom says, "Uh, oh, here we go!") . . . You take mild swipes at the counseling profession such as, 'I hear people are still trying to find out whether counseling works.' Or at me, like, 'Uh, oh, Mr. Counselor is getting a little hot under the collar.' I've ignored these remarks, but in ignoring them I've become a kind of accomplice in your behavior. Sometimes we seem to be working like a team. Other times you ask me to make decisions for you, you know, the 'Just Tell Me' routine. I've let myself get on edge with you, and that's not helping us at all. . . . (Tom says, "You want to call my games, huh?") . . . Tom, I guess that I prefer that we not even play games with each other. Perhaps we could spend a little time re-starting our relationship."

1. Critique the approach this counselor takes. What would you change?
2. Share your critique with a learning partner.

72

3. Consider the following situations and write out an immediacy response that helps the client challenge unhelpful perspectives and actions.
4. Share your responses with a learning partner and give feedback to each other. Together come up with a more effective immediacy response.

1. **The situation.** The client is a person of the opposite sex. You have had several sessions with this person. It has become evident that the person is attracted to you and has begun to make thinly disguised overtures for more intimacy. The person finds you both socially and sexually attractive. Some of the overtures have sexual overtones.

Immediacy response.

_____

_____

_____

_____

_____

_____

_____

_____

2. **The situation.** In the first session you and the client, a relatively successful businessman, 40, have discussed the issue of fees. At that time you mentioned that it is difficult for you to talk about money, but you finally settled on a fee at the modest end of the going rates. He told you that he thought that the fee was "more than fair." However, during the last few sessions he has been dropping hints about how expensive this venture is proving to be. He talks about getting finished as quickly as possible and intimates that is your responsibility. You, who thought that the money issue had been resolved, find it still very much alive.

Immediacy response.

_____

_____

_____

_____

_____

Supplemental exercises are found in Appendix Two.

EXERCISE 34: IMMEDIACY WITH THE OTHER MEMBERS OF YOUR TRAINING GROUP

This exercise makes sense only if you are a member of an experiential training group.

1. If the group is large, divide up into subgroups of about four trainees per group.
2. Read the example below.
3. On separate paper, write out a statement of immediacy for the members of your group. Imagine yourself in a face-to-face situation with each member successively. Deal with real issues that pertain to the training sessions, interactional style, and so forth.
4. In a round robin, share with each of the other members of the group the statement you have written for him or her.
5. The person listening to the immediacy statement should reply with empathy, making sure that he or she has heard the statement correctly.
6. Listen to the immediacy statement the other person has for you and then reply with empathy.
7. Finally, discuss for a few minutes the quality of your relationship with each other in the training group.
8. Continue with the round robin until each person has had the opportunity to share an immediacy statement with every other member.

Example: Trainee A to Trainee B. "I notice that you and I have relatively little interaction in the group. You give me little feedback; I give you little feedback. It's almost as if there is some kind of conspiracy of non-interaction between us. I like you and the way you act in the group. For instance, I like the way you invite others to challenge themselves. You do it carefully but without any apology. I think I refrain from giving you feedback, at least negative feedback, because I don't want to alienate you. I do little to make contact with you. I have a hunch that you'd like to talk to me more than you do, but it's just a hunch. I'd like to hear your side of our story, or non-story, as the case might be."

With a learning partner, identify the elements of immediacy (self-disclosure, challenge, invitation) in this example. Then move on to the exercise.

5. SUMMARIZING: FOCUS AND CHALLENGE

At a number of points throughout the helping process it is useful for helpers to summarize or to have clients summarize the principal points of their interactions. This places clients under pressure to focus and move on. Since summarizing places pressure on clients to "move forward," it is often a form of challenge.

EXERCISE 35: SUMMARIZING AS A WAY OF PROVIDING FOCUS

This exercise assumes that trainees have been using the skills and methods of Steps I-A and I-B to help one another.

1. The total training group is divided into subgroups of three.
2. There are three roles in each subgroup: helper, client, and observer.
3. The helper spends about eight minutes counseling the client. The client should continue to explore one of the problem areas he or she has chosen to deal with in the training group.
4. At the end of four minutes, the helper summarizes the principal points of the interaction. Helpers should try to make the summary both accurate and concise. The helper can draw on past interactions if he or she is counseling a "client" whom he or she has counseled before. At the end of eight minutes or so, the helper should engage in a second summary.
5. At the end of each summary, the helper should ask the client to draw some sort of implication or conclusion from the summary. That is, the client is asked to take the next step.
6. Then both observer and client give the helper feedback as to the accuracy and the helpfulness of the summary. The summary is helpful if it moves the client toward problem clarification, goal setting, and action.
7. This process is repeated until each person in the subgroup has had an opportunity to play each role.

Example 1: It would be too cumbersome to print ten minutes of dialogue here, but consider this brief outline of a case. A young man, 22, has been talking about some developmental issues. One of his concerns is that he sees himself as relating to women poorly. One side of his face is scarred from a fire that occurred two years previous to the counseling session. He has made some previous remarks about the difficulties he has relating to women. After four minutes of interaction, the helper summarizes:

*   **Helper.** "Dave, let me see if I have the main points you've been making. First, because of the scars, you think you turn women off before you even get to talk with them. The second point, and I have to make sure that this is what you are saying, is that your initial approach to women is cautious, or cynical, or maybe even subtly hostile since you've come to expect rejection somewhat automatically."
*   **Dave.** "Yeah, but now that you've put it all together, I am not so sure that it's all that subtle."
*   **Helper.** "You also said that the women you meet are cautious with you. Some might see you as 'mean.' Some steer clear of you because they see you as a kind of 'difficult person.' What closes the circle is that you take their caution or aloofness as their being turned off by your physical appearance."
*   **Dave.** "I don't like to hear it that way, but that's what I've been saying."
*   **Helper.** "If these points are fairly accurate, I wonder what implication you might see in them."
*   **Dave.** "I'm the one that rejects me because of my face. Nothing's going to get better until I do something about that."

Note that the client draws an implication from the summary ("I am the primary one who rejects me") and moves on to some minimal declaration of intent ("I need to change this").

Example 2: A woman, 47, has been talking about her behavior in the training group. She feels that she is quite nonassertive and that this stands in the way of being an effective helper. She and her helper explore this theme for a few minutes and then the helper gives this summary:

- **Helper.** "I'd like to take a moment to pull together the main points of our conversation. You're convinced that the ability to 'intrude' reasonably into the life of the client is essential for you as a helper. However, this simply has not been part of your normal interpersonal style. If anything, you are too hesitant to place demands on anyone. When you take the role of helper in training sessions, you feel awkward using even basic empathy and even more awkward using probes. As a result, you let your clients ramble and their problems remain unfocused. Outside training sessions you still see yourself as quite passive, except now you're much more aware of it. If this is more or less accurate, what implication might you draw from it?"
- **Client.** "When I hear it all put together like that, my immediate reaction is to say that I shouldn't try to be a counselor. But I think I would be selling myself short. No matter what my career will be, I can't keep on being such a hesitant person. I have to learn how to take risks."

After each trainee gives his or her summary and elicits some reaction from the client, the feedback from the client and the observer should center on the accuracy and the usefulness of the summary, not on a further exploration of the client's problem. Recall that feedback is most effective when it is clear, concise, behavioral, and nonpunitive.

Section 10
STEP I-C: LEVERAGE — HELPING CLIENTS
WORK ON THE RIGHT THINGS

The exercises in this section deal with helping clients work on issues that will make a difference in their lives, that is, issues that have "leverage." Counseling time is too precious to waste on issues that do not add value and make a difference in the client's life. The final exercise in this section helps you pull together all you have learned about Stage I of the helping process.

EXERCISE 36: MYSELF AS A DECISION MAKER

Since, as a counselor, you will be helping clients to make decisions (which, of course, is not the same as making decisions for them), it is useful to take a look at yourself as a decision maker.

1. Read the first part of Chapter Ten in *The Skilled Helper*, Leverage — Helping Clients Work on the Right Things.
2. With a learning partner, discuss the difference between "rational" decision making and the "shadow side" of decision making.
3. On your own, review the way you go about making decisions. Then, on a separate piece of paper, write a short description of your decision-making style. Include the "shadow side" of your decision-making style.

4. Swap descriptions with a learning partner. Read them. Then discuss how you might improve your style.

EXERCISE 37: SCREENING

Clients may need help in determining whether their issues are important enough to bring to a helper in the first place. This is called "screening."

1. Read these two case summaries with a view to discussing them with your fellow trainees.

- **Case 1**: Lila comes to a counselor for help. In telling her story, she says that she has occasional headaches and a few arguments with her husband, Lance. Doctors have told her that there is nothing wrong with her physically. The helper, using empathy, probing, and an occasional challenge, discovers that Lila is not understating the nature of her concerns and that there are no further hidden issues. She does discover that Lila and Lance have no children, that Lance works rather long hours in his new job, and that Lila is mainly a householder in their small condominium and has few outside interests.

- **Case 2**: Ray, 41, is a middle manager in a manufacturing company located in a large city. He goes to see a counselor with a somewhat complex story. He is bored with his job; his marriage is lifeless; he has poor rapport with his two teenage children, one of whom is having trouble with drugs; he is drinking heavily; his self-esteem is low; he has begun to steal things, small things, not because he needs them but because he gets a kick out of it. He tells his story in a rather disjointed way, skipping around from one problem area to another. He is a talented, personable, engaging man who seems to be adrift in life. He does not show any symptoms of severe psychiatric illness. He does experience a great deal of uneasiness in talking about himself. This is his first visit to a helper.

2. Discuss these two cases in terms of the material in the text on screening. Consider these questions:

- How would you approach the woman in Case 1?
- Under what conditions would you be willing to work with her?
- How would you approach the man in Case 2?
- In what general ways does this case differ from Case 1?
- What would your concerns be in working with the man in Case 2?

EXERCISE 38: CHOOSING ISSUES THAT MAKE A DIFFERENCE

Often enough, the stories clients tell are quite complex. And so they may need your help in deciding which issues to work on first and which merit substantial attention. In this exercise you are asked to review the personal concerns and problems you have identified in doing the exercises in this manual.

1. Briefly list about ten concerns or unexploited opportunities you have discovered in doing the assessment exercises up to this point.

_____

_____

_____

_____

_____

_____

_____

_____

2. Now do some screening. Put a line through those that you would probably not bring to a counselor because they are not that important or because you believe that you could handle them easily if you wanted.

3. Review the rest in the light of the following "leverage" criteria taken from the text.

- a. If there is a crisis, first help the client manage the crisis.
- b. Begin with the problem that seems to be causing pain for the client.
- c. Begin with issues the client sees as important and is motivated to work on.
- d. Begin with some manageable sub-problem of a larger problem situation.
- e. Begin with a problem that, if handled, will lead to some kind of general improvement in the client's condition.
- f. Focus on a problem for which the benefits will outweigh the costs.

4. Finally, using these criteria, evaluate the items remaining on your list. Next to each concern place the letters of the applicable "leverage" criteria listed.

Example: Gino is a trainee in a clinical psychology program. Here is one of the concerns on his list: "I am very inconsistent in the way I deal with people. For instance, some of my friends see me as fickle, I blow hot and cold. One friend told me that whenever he sees me, he's not sure which Gino he will meet. In fact, I seem to be inconsistent in other areas of life. Sometimes I give myself wholeheartedly to my studies, sometimes I couldn't care less." Gino believes that the following criteria apply: c, e, f.

5. Choose two issues that have high-leverage value for you. These are issues, concerns, opportunities, or problems, that, if pursued, would make a difference in your life.

6. Explain, in terms of the criteria outlined above or any further criteria not listed there, why you think that each has leverage. What is the "bang for the buck" of each?

7. Share one of these with a learning partner. Discuss what it would take to get you to invest time and energy in working with this problem and/or opportunity.

PULLING STAGE I TOGETHER

The following exercise asks you to pull together what you have learned about Stage I of the helping process and apply it to yourself.

EXERCISE 39: COUNSELING YOURSELF: AN EXERCISE IN STAGE I

At this stage, you have developed an overview of the helping model and, through reading and doing the exercises in this manual, developed an understanding and behavioral "feel" for Stage I. In this exercise you are asked to carry on a dialogue with yourself in writing. Choose a problematic area of your life, one that is relevant to your hoped-for success as a helper. First use empathy, probes, and challenges to help yourself tell your story. Choose a high-leverage issue for exploration. Work at clarifying it in terms of specific experiences, behaviors, and feelings. The dialogue should include probes for and challenges to engage in problem-managing and opportunity-developing action.

Example: This example comes from the experience of Cormack, a man in a master's degree program in counseling psychology. Here, then, is Cormack's dialogue with himself.

His Initial Story. "To be frank, I have a number of misgivings about becoming a counselor. A number of things are turning me off. For instance, one of my instructors this past semester was an arrogant guy. I kept saying to myself, 'Is this what these psychology programs produce? Could this guy really help anyone?' I also find the program much too theoretical. In a 'Theories of Counseling and Psychotherapy' course we never did anything, not even discuss the practical implications of these theories. And so they remained just that — theories. I'm very disappointed. I'm about to go into my second year, but I've got serious reservations. From what others tell me, the program gets a bit more practical, but not enough. There's a practicum experience at the end of the program, but I need more hands-on work now. So I've started working at a halfway house for people discharged from mental hospitals. But that's not working out the way I expected either. There's something about this whole helping business that is making me think twice about myself and about the profession."

Response to self. "All of this adds up to the fact that the helping profession, at least from your experience, is not what it's cracked up to be. The program and some of the instructors in it have left you quite disappointed. Of all these issues, which one or ones hits you the hardest?"

Self. "Hmm. It's hard to say, but I think the halfway house bothers me most. Because that's not theoretical stuff. That's real stuff out there."

Response to self. "That's a place where real helping should be taking place. But you've got misgivings about what's going on there."

Self. "Yes, two sets of misgivings. One set about me and one about the place."

Response to self. "Which set do you want to explore?"

Self. "I feel I have to explore both, but I'll start with myself. I feel so ill-prepared. What's in the lectures and books seems so distant from the realities of the halfway house. For instance, the other day one of the residents there began yelling at me when we were passing in the hallway. She hit me a few times and then ran off screaming that I was after her. I felt so incompetent. I felt guilty."

Response to self. "That sounds pretty upsetting. You just weren't prepared for it. I'm wondering whether the more practical part of the counseling program you're going into would better prepare you for that kind of reality."

Self. "It could be. I may be doing myself in by jumping ahead of myself."

Response to self. "But you still have reservations about the effectiveness of the halfway house."

Self. "I wasn't ready for what I found there. I've been there a couple of months. No one has really helped me learn the ropes. I don't have an official supervisor. I see all sorts of people with problems and help when I can."

Response to self. "You just don't feel prepared, and they don't do much to help you. So you feel inadequate. You also seem to be basing your judgments about the adequacy of helper-training programs and helping facilities on this training program and on the halfway house."

Self. "That's a good point. I'm making the assumption that both should be high-quality places. As far as I can tell, they're not. I guess I have to work on myself first. But that's why I went to the halfway house in the first place. I'm an independent person, but I'm too much on my own there. In a sense, I'm trusted, but, since I don't get much supervision, I have to go on my own instincts, and I'm not sure they're always right."

Response to self. "There's some comfort in being trusted, but without supervision you still have a what-am-I-doing-here feeling."

Self. "There are many times when I ask myself just that, 'What are you doing here?' I provide day-to-day services for a lot of people. I listen to them. I take them places, like to the doctor. I get them to participate in conversations or games and things like that. But it seems that I'm always just meeting the needs of the moment. I'm not sure what the long-range goals of the place are and if anyone, including me, is contributing to them in any way."

Response to self. "You get some satisfaction in providing the services you do, but this lack of overall purpose or direction for yourself and the institution is frustrating. I'm not sure what you're doing about all of this, either at the halfway house or in the counseling program."

Self. "I'm letting myself get frustrated, irritated, and depressed. I'm down on myself and down on the people who run the house. It's a day-to-day operation that sometimes seems to be a fly-by-night venture. See! There I go. You're right. I'm not doing anything to handle my frustrations. I am doing something to try to better myself, that is, to make myself a better helper, but this is the first time I've expressed myself about the halfway house or about the counseling program."

Response to self. "So this is really the first time you've stopped to take a critical look at yourself as a potential helper and the settings in which helping takes place. And what you see is depressing.

Self. "Absolutely."

Response to self. "I wonder how fair you're being to yourself, to the profession, and even to the halfway house."

Self. (Pause) "Well, that's a point. I'm speaking — and making judgments — out of a great deal of frustration. I don't want to go off half cocked. But I could share my concerns about the psychology program with one of the instructors. She teaches the second-year trainees. She gave a very practical talk. I could also see whether my concerns are shared by my classmates. I could also talk to the second-year students to get a feeling for how practical next year might be."

Response to self. "Give yourself a chance to get a more balanced perspective."

Self. "Right. I need to do something instead of just brooding and complaining. But there's still the halfway house. I don't want to come across as the wet-behind-the-ears critic."

Response to self. "I'm not sure whether you're assuming that particular halfway house should be a state-of-the-art facility?"

Self. "Touché! . . . . Wow, I am basing a lot of my feelings about the profession on that place."

Response to self. "Who out there might you trust to help you get a better perspective?"

Self. "No one I can think of. But there is a consultant who shows up once in a while. He runs staff meetings in a very practical way. Maybe I can get hold of him and get a wider picture."

Response to self. "So, overall?"

Self. "I've got some work to do before I rush to judgment. I like the fact that I want the people and the institutions in the profession to be competent. Deep down I think that I'd make a good helper. I say to myself, 'You're all right; you're trying to do what is right.' Also, I need to challenge my idealism. I spend too much time grieving over what is happening at school and at the halfway house. I need to figure out how to turn minuses into pluses."

1. Review this trainee's responses to himself with a learning partner. What kinds of responses did he use? How would you evaluate their quality? Did they get him someplace? Describe the movement he made during the session. What responses would you have changed?

2. Choose a problematic area that is important to you and on separate sheets of paper engage in the same kind of dialogue with yourself. Tell your story briefly, choose a high-leverage issue, and clarify it in terms of specific experiences, behaviors, and feelings. Stay within the steps of Stage I.

3. Exchange dialogues with your learning partner and engage in the same kind of critique outlined above.

HELPING CLIENTS DEVELOP PROGRAMS FOR CONSTRUCTIVE CHANGE

In many ways Part Four, which includes Stages II and III, is the most important part of the helping model. It is here that counselors help clients develop and implement programs for constructive change. The payoff for identifying and clarifying both problem situations and unused opportunities lies in doing something about them. This skills that both helpers and clients use to do precisely this — engage in constructive change — are reviewed and illustrated in Stages II and III. Chapter Eleven in *The Skilled Helper*, Perspectives and Skills for Constructing a Better Future, is an introduction to Stages II and III of the helping model. Since no exercises are based on that chapter, there is no Section Eleven in this manual.

STAGE II
HELPING CLIENTS CREATE A BETTER FUTURE

Stage II focuses on a better future, the client's preferred scenario. Problems can make clients feel hemmed in and closed off. To a greater or lesser extent they have no future, or the future they have looks troubled. The steps of Stage II outline three ways in which helpers can be with their clients at the service of exploring and developing this better future.

Step II-A: Helping clients identify *possibilities* for a better future.
Step II-B: Helping clients choose specific preferred-scenario *goals*.
Step II-C: Helping clients discover incentives for *commitment* to their goals.

The three steps of Stage II become clearer to clients when they are translated into everyday language. The questions below are asked, of course, against the background of some degree of understanding on the part of the client of the problem situation or of the unused opportunity.

Step II-A: What do you want? What do you need? What are some of the possibilities?
Step II-B: Given the possibilities, what do you really want? What are your choices?
Step II-C: What are you willing to pay for what you want?

Section 12
STEP II-A: A BETTER FUTURE
WHAT DO YOU WANT? — DISCOVERING POSSIBILITIES

Effective helping is related to the use of imagination. In this step you are asked to help yourself and clients develop a vision of a better future. Once clients understand the nature of the problem situation, they need to ask themselves, "What do I want? What would my situation look like if it were better, at least a little bit better?"

EXERCISE 40: DEVELOPING POSSIBILITIES FOR A BETTER FUTURE

In this exercise you are asked to use your imagination to build a better "future" for yourself as a way of preparing you to help others develop new scenarios.

Example: Since most students do not operate at 100% efficiency, there is usually room for improvement in the area of learning. Luisa, a junior beginning her third year of college, is dissatisfied with the way she goes about learning. She decides to use her imagination to invent a new study scenario. She brainstorms possibilities for a better study future, that is, goals that would constitute her new learning style. She says to herself: "In my role as student or learner, what do I want? Let me brainstorm the possibilities."

- I will not be studying for grades, but studying to learn. Paradoxically this might help my grades, but I will not be putting in extra effort just to raise a B to an A.
- I will be a better contributor in class, not in the sense that I will be trying to make a good impression on my teachers. I will do whatever I need to do to learn. This may mean placing more demands on teachers to clarify points, making more contributions, and involving myself in discussions with peers.
- I will have in place a more constructive approach to writing papers. For instance, once a paper is assigned, I will start a file on the topic and collect ideas, quotes, and data as I go along. Then, when it comes to writing the paper, I will not have to try to create something out of nothing at the last moment. I assume this will help me feel better about the paper and about myself.
- I will be reading more broadly in the area of my major, psychology, not just the articles and books assigned but also in the areas of my interest. I will let my desire to know drive my learning.

Luisa goes on to draw up a fairly extensive list of the patterns of behavior that might have a place in her new scenario, that is, her new approach to learning. Only when she has an extensive list does she address the task of evaluating and choosing the actual goals that will constitute her new learning style.

1. Read the material on brainstorming in Chapter Eleven.
2. Read Chapter Twelve in *The Skilled Helper*.
3. Choose two problem areas or undeveloped opportunities that you have been working on.
4. Like Luisa, brainstorm possibilities for a better future in these two areas.
5. Use probes such as the following to help yourself develop preferred-scenario possibilities:
- Here's what I need . . . .
- Here's what I want . . . .
- Here are some items on my wish list . . . .
  - When I'm finished I will have . . . .
  - There will be . . . .
  - I will have in place . . . .
  - I will consistently be . . . .
  - There will be more of . . . .
  - There will be less of . . . .

a. Problem/opportunity area # 1.

_____

_____

b. Possibilities for a better future.

_____

_____

_____

_____

_____

_____

a. Problem/opportunity area # 2.

_____

_____

b. Possibilities for a better future.

_____

_____

_____

_____

_____

_____

6. When you are finished, share the fruits of your brainstorming with a learning partner.
7. Help each other add three more items to the list.

EXERCISE 41: HELPING ANOTHER DEVELOP POSSIBILITIES FOR A BETTER FUTURE

In this exercise, you are asked to help one of the other members of your group develop possibilities for a better future.

1. Choose a partner for this exercise.
2. One partner takes the role of client and the other the role of the helper.
3. The client gives a summary of one of the problem situations focused on in the previous exercise.
4. The client then shares his or her list of new-scenario possibilities (the ones developed in the previous exercise).
5. The helper, using empathy, probes, and challenges, helps the client clarify and add to the items already on the list. Overall, the helper helps the client tap into his or her imagination more fully.

Example: In the following example, Geraldo, a junior in college majoring in business studies, has given a summary of the problem situation and his list of preferred-scenario possibilities. Overall he wants a more balanced lifestyle. His uncle, who runs a small business, has offered him a kind of internship in the family business.   While such an opportunity fits perfectly with Geraldo's career plans, he has done nothing to develop it.   Other things such as studies, intramural sports, and a rather substantial social life have crowded it out. Given his ambitions, Geraldo's priorities are out of line. Trish, his helper, challenges his list.

Trish. "There seems to be a contradiction. Very few of the possibilities you have outlined for a more balanced lifestyle relate to your uncle's offer. Yet earlier you said you wanted to develop the serious side of things a bit more."
Geraldo. "I don't want work to consume my life. I want a balanced life, not like some of those guys who never come home from work."
Trish. "In what way is your life out of balance right now?"
Geraldo. "Well, there's a bit too much play, I suppose."
Trish. "If that's the case, let's brainstorm more work-related possibilities. You can take care of the balance when you actually set your agenda. Spell out more possibilities that relate to your uncle's offer. You already said he's not going to push you into anything."
Geraldo. "Let's see. I'd be putting in 10 to 15 hours a week at my uncle's business."
Trish. "What would that look like?"
Geraldo. "Well, I don't know what my uncle has in mind."
Trish. "Well, what do you want? Name some of the things you'd like to get out of that kind of internship."

Geraldo. "Without being a nuisance, I'd like to make my uncle's place something like a lab for business studies. For instance, I've read about corporate culture and the way it can strangle a business. I'd like to see if there's a culture in my uncle's business and what it's like. There are only about a hundred people working there. So it's lab size, as it were."

Trish. "That would make it more than just work, but a place to learn a lot of practical things."

Geraldo. "I'd also like to learn something about the finance part of his business — where the money comes from, what kind of debt he carries, cash flow — all those things that are still too theoretical for me in the courses I'm taking."

Trish. "A way of bringing the texts alive. What else?"

Geraldo. "I'd like to learn something about the role of the manager. Especially in directing the work of others. I'm not sure what a manager really does with his day."

Trish. "That sounds practical."

Geraldo. "Yeah. It would certainly send me back to the books in a different way. I saw a couple of articles on new approaches to strategic management and on employee empowerment in the *Harvard Business Review*. But they were just theory. My uncle's place would help me turn theory into practice."

Trish. "So there could be a lot of synergy between the internship and your studies."

Geraldo. "Much more than I realized."

The dialogue goes on in that vein. Geraldo, with the help of Trish, develops not only a lot of possibilities for a better future but also much more enthusiasm about his uncle's project. Geraldo gives Trish high marks for empathy, probing, and challenge.

6. After ten minutes, the helper gets feedback from the client as to how useful he or she has been.

7. They then switch roles and repeat the process.

Section 13
STEPS II-B AND II-C:
WHAT DO YOU REALLY WANT — CHOICES AND COMMITMENT

This section deals with two key tasks: (1) helping clients set problem-managing and opportunity-developing goals, and (2) helping them find ways of committing themselves to these goals.

STEP II-B: FORMULATING GOALS
What Do You Really Want?

Once clients have brainstormed possibilities for a better future, they need to make some choices. Another way of putting this is that they need to set goals as part of an agenda for constructive change. For instance, Geraldo, the college student seen in Section 12, after brainstorming a number of possibilities for structuring his internship in his uncle's company, chooses four of the possibilities and makes these his agenda. He shares them with his uncle and, after some negotiation, comes up with a revised package they can both live with.

EXERCISE 42: TURNING POSSIBILITIES INTO GOALS

In this exercise you are asked to do what Geraldo had to do — choose several preferred-scenario possibilities as the first step in crafting an agenda. Possibilities should be chosen because they will best help you manage some problem situation or develop some opportunity. First, read the following case.

The Case. Vanessa, 46, has been divorced for about a year. She has done little to restructure her life and is still in the doldrums. At the urging of a friend, she sees a counselor. With her help, she brainstorms a range of possibilities for a better future around the theme of "my new life as a single person." She is currently a salesperson in the women's apparel department of a moderately upscale store. She lives in the house that was part of the divorce settlement. She has no children. When she finally decided that she wanted children, it was too late. There was some discussion about adopting a child, but it didn't get very far. The marriage was disintegrating. Her visits to the counselor have reawakened a desire to take charge of her life and not just let it happen. She let her marriage happen, and it fell apart. She is not filled with anger at her former husband. If anything, she's a bit too down on herself. Her grieving is filled with self-recrimination.

A helper asked her, "What do you want now that the divorce is final? What would you like your post-divorce life to look like? What would you want to see in place?" Vanessa brainstormed the following possibilities:

- A job related to the fashion industry; maybe a career later.
- A small condo that will not need much maintenance on my part instead of the house.
- The elimination of poor-me attitudes as part of a much more creative outlook on self and life.
- The elimination of waiting around for things to happen.
- The development of a social life. For the time being, a range of friends rather than potential husbands. A couple of good women friends.
- Getting into physical shape.
- A hobby or avocation that I could get lost in. Something with substance.
- Some sort of volunteer work. With children, if possible.
- Some religion-related activities, not necessarily established-church related. Something that deals with the "bigger" questions of life.
- Some real "grieving" work over the divorce instead of all the self-recrimination.
- Resetting my relationship with my mother (who strongly disapproved of the divorce).
- Getting over a deep-seated fear that my life is going to be bland, if not actually bleak.
- Possibly some involvement with politics.

1. Mustering all the empathy within you, try to read the list from her point of view (even though you hardly know her).
2. Since Vanessa cannot possibly do all of these at once, she has to make some choices. Put yourself in her place. Which items would you include in your agenda? On separate pages of paper, indicate the items and your reasons for choosing each. In what way would the agenda item, if accomplished, help manage the overall problem situation and develop some opportunities?
3. Share your "package" with a learning partner. Discuss the differences in and the reasons for the choices. Note that there is no one right package.

EXERCISE 43: MAKING GOALS SPECIFIC

This exercise assumes that you are familiar with the material in Chapter Twelve. Many clients are more likely to pursue goals if they are clear and specific. Your job here is to move from the general to the specific.

Example: Tom, 42, and his wife, Carol, 39, have been talking to a counselor about how poorly they relate to each other. They have agreed to stop blaming each other, have explored their own behavior in concrete ways, have developed a variety of new perspectives on themselves as both individuals and spouses, and now want to do something about what they have learned.

Without having specific information about the issues Tom and Carol have discussed, use your imagination to come up with three levels of concreteness in a goal-shaping process that might apply to their situation. That is, move from some good intention to a broad goal and then turn the broad goal into one or more specific goals. Ultimate specific goals can be measured in some way. Obviously, in an actual counseling situation, you would be helping them shape their own goals. This exercise deals with goals (*what* needs to be put in place), not with strategies (*how* any specific goal is to be accomplished). What follows is Tom and Carol's first shot at moving from a vague to a more specific goal.

- **Good Intention.** "We've got to do something about saving our marriage because it is worth saving."
- **Broad Goal.** "We'd like to improve the quality of the time we spend together at home."
- **Specific Goal.** "We want to have better give-and-take, problem-solving conversations with each other."
- **Measurable Goal.** "Over the next month, we want to cut in half the number of times our so-called discussions turn into arguments or out-and-out fights."

Note that each level becomes more specific in some way. Note, too, that their specific goal is negative; they do not say what they'd like to put in place of fighting. This might be a flaw in the goal-setting process. Now do the same with the situations listed below.

1. Use your imagination to develop aims and goals. Get inside the client's mind and try to think the way the client might think.
2. Move from a vague statement of good intention to a specific goal in each case.

1. Linda W., 68, is dying of cancer. She has been talking to a pastoral counselor about her dying. One of her principal concerns is that her husband does not talk to her about her impending death. She has a variety of feelings about dying that well up from time to time such as disbelief, fear, resentment, anger, and even peace and resignation. She also has thoughts about life and death that she has never had before and has never shared with anyone.

Statement of good intention.

_____

_____

Broad goal.

_____

_____

Specific goal.

_____

_____

How is it to be measured? How will we know that it has been accomplished?

_____

_____

2. Troy, 30, has been discussing the stress he has been experiencing during this transitional year of his life. Part of the stress relates to his job. He has been working as an accountant with a large firm for the past five years. He makes a decent salary, but he is more and more dissatisfied with the kind of work he is doing. He finds accounting predictable and boring. He doesn't feel that there's much chance for advancement in this company. Many of his associates are much more ambitious than he is.

Statement of good intention.

_____

_____

Broad goal.

_____

_____

Specific goal.

_____

_____

How is it to be measured? How will we know that it has been accomplished?

_____

_____

3. Joan, 32, is married and has two small children. Her husband has left her, and she has no idea where he is. She has no relatives in the city and only a few acquaintances. She is talking to a counselor in a local community center about her plight. Since her husband was the breadwinner, she now has no income and no savings on which to draw.

Statement of good intention.

_____

_____

Broad goal.

_____

_____

Specific goal.

_____

_____

How is it to be measured? How will we know that it has been accomplished?

_____

_____

Supplemental exercises are found in Appendix Two.

EXERCISE 44: SHAPING VIABLE GOALS FOR YOURSELF

Counselors can add value by using the communication skills and helping methods and techniques discussed up to this point to help clients choose, craft, shape, and develop their goals. Goals are specific statements about what clients want and need. The goals that either emerge or that are explicitly set by clients in fashioning problem-managing and opportunity-developing programs for constructive change should have, at least eventually, certain characteristics in order to be _viable_ goals. They should be:

- **outcomes** rather than activities.
- **specific** enough to be verifiable and to drive action.
- **substantive** and challenging.
- **realistic** and sustainable.

- in keeping with the client's **values**.
- set in a reasonable **time frame**.

From a practical point of view, these characteristics can be seen as "tools" that you can use to help clients shape their goals.

1. Take these possibilities and see whether they are statements of intent, general aims, more specific aims, or specific goals. If they are not specific goals, shape them so that they are.

Example: Jeff, a trainee in a counseling psychology program, has been concerned that he does not have the kind of assertiveness that he now believes helpers need in order to be effective in consultations with clients. He is specifically concerned about the quality of his participation in the training group. First he translates his intention to be more active into a much more specific goal:

- **Good Intention.** "I need to be more assertive if I expect to be an effective helper."
- **Broad Goal.** "I want to take more initiative in this training group."
- **Specific Goal.** "In our group sessions, when there is relatively little structure, I want to speak up without being asked to do so. Specifically, I want to practice empathy much more."
- **Measurable Goal.** "In the next training session, without being asked to do so, I will respond to what others say with empathy. During our two-hour meeting, I will respond at least ten times with basic empathy when other members talk about themselves."

Jeff then evaluates his specific and measurable goals by applying the criteria for viable goals:

- **Outcome.** "This goal is stated as an outcome, that is, a pattern of assertive responding to be accomplished."
- **Specific.** "It is clear. I can actually picture myself using empathy. It is quite easy to verify whether I have accomplished my goal or not. I can get feedback from the other members of the group and from the trainer. It might even be too specific because I have limited it to that one session."
- **Substantive.** "Responding with empathy with some frequency will help me develop, at least in part, the kind of assertiveness called for in helping. In this sense, my goal is a real step forward."
- **Realistic.** "I know how to respond with empathy. I think I do it well, when I do it. I can summon up the guts needed to use the skill in the group."
- **Values.** "This goal is in keeping with my values of being a good listener and of taking responsibility for myself as a trainee. The value of being an assertive, proactive helper, discussed in Chapter Three of *The Skilled Helper*, is new to me. While I like the idea, I need to work at making this value my own. My goal is a step in this direction."
- **Time frame.** "The time frame should refer to my developing a more assertive pattern of responding with empathy and not focus on just the next training session. By the end of the semester, I will be consistently participating the way I intend to in the next training session."

2. With a learning partner, critique Jeff's change program.
3. Finally, name two goals you have developed so far to become a more effective helper. Do what Jeff did above for each goal.

Goal # 1.

_____

_____

_____

_____

Goal # 2.

_____

_____

_____

_____

4. Share your goals with and get feedback from a learning partner.

EXERCISE 45: HELPING CLIENTS SET VIABLE GOALS

Example: Tom and Carol's specific goal is: "Over the next month, we want to cut in half the number of times our conversations turn into arguments or out-and-out fights."

- **Outcome**. "Number of fights _decreased_" is an accomplishment. A new pattern of behavior would be _in place_.
- **Specific**. It is behaviorally _clear_. They can get a picture of themselves not arguing or fighting. It probably would help if they were more specific, that is, if they actually talked about what an aborted fight or argument would look like. Moreover, it is not clear what they will be doing instead of fighting, such as substituting some kind of problem-solving or negotiation dialogue. Furthermore, since they have some idea of how often they fight per day or week, they can verify whether the number of fights has decreased. At this point, just some kind of counting might be in order. The goal says nothing about the intensity or viciousness of fights. Perhaps that should have been taken into account.
- **Substance.** It makes sense to suppose that a decrease in the number of arguments or fights will contribute substantially to the betterment of their relationship. However, stopping fighting leaves a void. They had better talk about the void. There is also something contrived about this goal. People don't usually talk about cutting the number of their fights in half.
- **Realism.** Tom cannot control his wife's behavior, but he can control his own. The same can be said of Carol. The assumption is that both of them have the self-management skills and the emotional resources to back off from fights. They don't have any experience in cutting

92

off fights before they begin. Furthermore, neither has something that they are going to put in place of fighting. In what ways does "number of fights decreased" appeal to both of them? There are some problems with realism here.

- **Values.** They both espouse give-and-take and fairness in their relationship, but they have been poor at delivering what they promise. They need to talk to one another more about what they need and want from the marriage. Their values right now are not clear.
- **Time frame.** "Over the next month" is the time frame. They have to determine whether this is a realistic time frame or whether they should move more to some kind of phasing out of unwanted behaviors (together with phasing in the behaviors that are to take the place of the unwanted ones).

Restated goal. In the light of the above analysis, Carol and Tom restate their goal. "We want to substantially reduce the number of arguments we have. We need more peace at home. Our goal is to put in place a process that will help us fight less. We've devised a time-out process. Anytime either of us sees things heating up, he or she can call time-out to discuss what's going on. A time-out session is not a who's-to-blame session. Rather we want to find out what the process is that leads to all the fighting. We've actually tried it a few times, and it works. We end up laughing at ourselves."

- Using the criteria above, critique the restated goal.
- Reshape the goal in the light of your critique. How would it read?

_____

_____

_____

_____

_____

1. Return to the three specific goals you have come up with in each of the cases in Exercise 43.
2. Critique each in terms of the principles outlined above.
3. If a goal does not meet these standards, use these principles as tools to shape it until it does. Then restate the goal.

EXERCISE 46: HELPING LEARNING PARTNERS SET VIABLE GOALS

In this exercise you are asked to act as a helper/consultant to a learning partner.

1. The total training group is to be divided up into smaller groups of three.
2. There are three roles: client, helper, and observer. Decide the order in which you will play each role.
3. The client summarizes some problem situation and then declares his or her **intent** to do something about the problem or some part of it.

a. A summary of your problem and/or unused opportunity situation.

_____ _____

_____ _____

_____ _____

b. Your statement of intent.

_____ _____

_____ _____

_____ _____

4. The helper, using empathy, probing, and challenging, helps the client move from this statement of intent to a specific problem-managing or opportunity-developing goal that has the characteristics listed above.

5. When the helper feels that he or she has fulfilled this task, the session is ended and both observer and client give feedback to the helper on his or her effectiveness.

6. Repeat the process until each person has played each role.

EXERCISE 47: RELATING GOAL CHOICE TO ACTION

Once clients state what they want, they need to move into action in order to get what they want. At this point there are two ways of looking at the relationship of goals to action, one formal and one informal.

- **Formal.** Stage III of the helping model — brainstorming action strategies, choosing the best package, and turning them into a plan for accomplishing goals — is the formal approach to action.
- **Informal.** Once clients get an idea of what they want, there is no reason why they cannot move immediately into action, that is, do *something* that will move them in the direction of their goals. These are the "little" actions that can precede the formal planning process. This exercise is about these little actions.

Example: Larry, a young man with AIDS, wants to accomplish a lot of things before he dies. His overall aim is to live until he dies and to live as fully as possible. Since his lifestyle has alienated many of the members of his rather conservative extended family, one of his goals is some kind of reintegration into the family. He wants to feel that he belongs inside the family. He wants to be accepted by his relatives as they currently accept one another. At this point he wants everyone to know that he has AIDS. Up to this point it has been somewhat of a secret. Larry knows that he can't force anyone to accept him, so his goal focuses on creating a climate or set of conditions where this kind of acceptance is possible. Once he has chosen this overall goal, he does the following immediately: First, he moves back into town and takes up residence in the family

94

home. His parents, while bewildered by his lifestyle, have always been accepting. They are glad that he has decided to come home. Second, he tickles the informal communication network. He makes it clear to a couple of his relatives who are still relatively close to him that he wants everyone to know that he has AIDS. Third, he starts going to church again. These three actions are done quickly, almost instinctively, and are not part of a formal plan. They are "little" actions that head him in the right direction, that is, toward reintegration into the family. He has not yet developed a more formal plan.

1. Reacquaint yourself with the case of Vanessa discussed in Exercise 42.
2. While she believes that it is essential for her to develop a better attitude about herself and life in general, she believes that it is not best to try to do this directly. Therefore, she chooses goals that will have a better attitude as a by-product.
- One goal centers around the development of a social life. For the time being, she wants a range of friends rather than potential husbands. She would especially like a couple of good women friends.
- A second goal revolves around religion in some sense of that term. She would like to have religion-related activities, not necessarily established-church related, as part of her life. Something that deals with the "bigger" questions of life.
- A third goal relates to physical well-being. The stress of the divorce has left her exhausted. She wants to build herself up physically and get into good physical shape.

3. If you were Vanessa and these were your goals, what are a few things you could do right away to get yourself moving in the right direction even before developing a formal plan to implement each goal?

Goal # 1.

_____

_____

_____

Goal # 2.

_____

_____

_____

Goal # 3.

_____

_____

_____

STEP II-C: COMMITMENT
What Are You Willing to Pay for What You Want?

Many of us choose goals that will help us manage problems and develop opportunities, but we do not explore them from the viewpoint of commitment. Just because goals are tied nicely to the original problem situation and initially are espoused by us does not mean that we are really committed to them nor that we will follow through. The work of discussing problems and setting goals to manage them is costly in terms of time, psychological effort, and expense. A directconsideration of commitment can raise the probability that clients will actually pursue these goals.

EXERCISE 48: REVIEWING THE COST/BENEFIT RATIO IN THE CHOICE OF GOALS

In most choices we make there are both benefits and costs. Commitment to a preferred-scenario goal often depends on a favorable cost/benefit ratio. Do the benefits outweigh the costs?

Example: In January, Helga, a married woman with two children, one a senior in college and one a sophomore, was told that she had an advanced case of cancer. She was also told that a rather rigorous series of chemotherapy treatments might prolong her life, but they would not save her. She desperately wanted to see her daughter graduate from college in June, so she opted for the treatments. Although she found them quite difficult, she buoyed herself up by the desire to be at the graduation. Although in a wheel chair, she was there for the graduation in June. When the doctor suggested that she could now face the inevitable with equanimity, she said: "But, doctor, in only two years my son will be graduating."

This is a striking example of a woman's deciding that the costs, however high, were outweighed by the benefits. Obviously, this is not always the case. This exercise gives you the opportunity to explore your goals from a cost/benefit perspective. Is it worth the effort? What's the payoff?

1. Divide up into pairs, with one partner acting as client, one as helper.
2. Help your partner review one of the goals of his or her agenda from a cost/benefit perspective. The helper is to use basic empathy, probing, and challenging to help his or her partner do this. Help your partner identify benefits and costs and do some kind of trade-off analysis such as the balance sheet technique (see the text).
3. Help you partner clearly state the incentives and payoffs that enable him or her to commit to the specified goals.
4. After the discussion, each is to get a new partner, change roles, and repeat the process.

EXERCISE 49: MANAGING YOUR COMMITMENT TO YOUR GOALS

In this exercise you are asked to review the goals you have chosen to manage some problem situation with a view to examining your commitment. It is not a question of challenging your goodwill. All of us, at one time or another, make commitments that are not right for us.

1. Review the problem situation you have been examining and the preferred-scenario goals you have established for yourself as a way of managing it or some part of it.

2. Review the material on choice and commitment in the text and then use the following questions to gauge your level of commitment:

- To what degree are you choosing this goal freely?
- Are your goals chosen from among a number of possibilities?
- How highly do you rate the appeal of your goals?
- Name any ways in which your goals do not appeal to you.
- What's pushing you to choose these goals?
- If any of your goals are imposed by others, rather than freely chosen, what incentives are there besides mere compliance?

3. With a learning partner, review your principal learnings from answering the above questions about two of your goals. Take turns. Use empathy, probes, and challenges to help one another explore levels of commitment.

4. If you have any hesitations about committing yourself to a goal, discuss these hesitations with your partner. Use the following questions in the discussion.

- What is your state of readiness for change in this area at this time?
- What difficulties do you experience in committing yourself to your agenda or any part of it?
- What stands in the way of your commitment?
- What can you do to get rid of the disincentives and overcome the obstacles?
- What can you do to increase your commitment?
- To what degree is it possible that your commitment is not a true commitment?
- In what ways can the agenda be reformulated to make it more appealing?
- In what ways does it make sense to step back from this problem or opportunity right now? To what degree is the timing poor?

5. Finally, reformulate your goals in terms of what you have learned from the dialogue.

The Restated Goal.

_____

_____

_____

_____

_____

_____

STAGE III: GETTING THERE —
HELPING CLIENTS IMPLEMENT THEIR GOALS

Stage III deals with **what** clients need to do in order to accomplish their problem-managing and opportunity-developing goals. In this stage, counselors help clients brainstorm different ways of accomplishing their goals, choose the strategies that best fit available resources, and draw up formal plans to accomplish goals.

Section 14
STEP III-A: STRATEGIES —
WHAT DO YOU NEED TO DO TO GET WHAT YOU WANT?

There is usually more than one way to accomplish a goal. However, clients often focus on a single strategy or just a few. The task of the counselor in Step III-A is to help clients discover a number of different routes to goal accomplishment. Clients tend to choose a better strategy or set of strategies if they choose from among a number of possibilities. Read Chapter Fourteen before doing the exercises in this section.

EXERCISE 50: BRAINSTORMING ACTION STRATEGIES FOR YOUR OWN GOALS

Brainstorming is a technique you can use to help yourself and your clients move beyond overly constricted thinking. Recall the rules of brainstorming:

- Encourage quantity. Deal with the quality of suggestions later.
- Do not criticize any suggestion. Merely record it.
- Combine suggestions to make new ones.
- Encourage wild possibilities, "One way to keep to my diet and lose weight is to have my mouth sewn up."
- When you feel you have said all you can say, put the list aside and come back to it later to try once more.

Example: Ira, a retired lawyer in training to be a counselor, is in a high-risk category for a heart attack: some of his relatives have died relatively early in life from heart attacks, he is overweight, he exercises very little, he is under a great deal of pressure in his job, and he smokes over a pack of cigarettes a day. One of his goals is to stop smoking within a month. With the help of a nurse practitioner friend, he comes up with the following list of strategies:

- just stop cold turkey.
- shame myself into it, "How can I be a helper if I engage in self-destructive practices such as smoking?"
- cut down, one less per day until zero is reached.
- get the doctor to prescribe the new nicotine patches.
- look at movies of people dying with lung cancer.
- pray for help from God to quit.

- use those progressive filters that gradually squeeze all taste from cigarettes.
- switch to a brand that doesn't taste good.
- switch to a brand that is so heavy in tars and nicotine that even I see it as too much.
- smoke constantly until I can't stand it anymore.
- let people know that I'm quitting.
- put an ad in the paper in which I commit myself to stopping.
- send a dollar for each cigarette smoked to a cause I don't believe in, for instance, the "other" political party.
- get hypnotized; through a variety of post-hypnotic suggestions have the craving for smoking lessened.
- pair smoking with painful electric shocks.
- take a pledge before my minister to stop smoking.
- join a group for professionals who want to stop smoking.
- visit the hospital and talk to people dying of lung cancer.
- if I buy cigarettes and have one or two, throw the rest away as soon as I come to my senses.
- hire someone to follow me around and make fun of me whenever I have a cigarette.
- have my hands put in casts so I can't hold a cigarette.
- don't allow myself to watch television on the days in which I have even one cigarette.
- reward myself with a week-end fishing trip once I have not smoked for two weeks.
- substitute chewing gum for smoking, starting first with nicotine-flavored gum.
- avoid friends who smoke.
- have a ceremony in which I ritually burn whatever cigarettes I have and commit myself to living without them.
- suck on hard candy made with one of the non-sugar sweeteners instead of smoking.
- give myself points each time I want to smoke a cigarette and don't; when I have saved up a number of points, reward myself with some kind of "luxury."

Note that Ira includes a number of wild possibilities in his brainstorming session.

1. Now do the same for two goals you have set for yourself in order to manage some problem situation or develop some opportunity. Make sure that the goal has been properly shaped according to the principles in Step II-B. Vague goals will yield vague strategies.

Goal # 1.

_____

_____

_____

_____

_____

On a separate page, like Ira, brainstorm ways of achieving this goal. Observe the brainstorming rules. When you think you have run out of possibilities, stop and return to the task later.

Goal # 2.

_____

_____

_____

_____

_____

Brainstorm ways of achieving this goal. Add wilder possibilities at the end.

2. After you have finished your list, take one of the goals and the brainstorming list, sit down with a learning partner, and see if, through interaction with him or her, you can expand your list. Be careful to follow the rules of brainstorming.
3. Switch roles. Through empathy, probing, and challenge, help your partner expand his or her list.
4. Keep both lists of brainstormed strategies. You will use them in an exercise in Step III-B.

EXERCISE 51: ACTION STRATEGIES: PUTTING YOURSELF IN THE CLIENT'S SHOES

Here are a number of cases in which the client has formulated a goal and needs help in determining how to accomplish it. You are asked to put yourself in the client's shoes and brainstorm action strategies that you yourself might think of using were you that particular client.

Example: Richard, 53, has been a very active person — career-wise, physically, socially, and intellectually. In fact, he has always prided himself on the balance he has been able to maintain in his life. However, an auto accident that was not his fault has left him a paraplegic. With the help of a counselor he has begun to manage the depression that almost inevitably follows such a tragedy. In the process of re-directing his life, he has set some goals. Since his job and his recreational activities involved a great deal of physical activity, a great deal of re-direction is called for.

One of his goals is to write a book called "The Book of Hope" about ordinary people who have creatively re-set their lives after some kind of tragedy. The book has two purposes. Since it would be partly autobiographical, it would be a kind of chronicle of his own re-direction efforts. This will help him commit himself to some of the grueling rehabilitation work that is in store for him. Second, since the book would also be about others struggling with their own tragedies, these people will be models for him. Richard has never published anything, so the "how" is more difficult. For him, writing the book is more important than publishing it. Therefore, the anxiety of finding a publisher is not part of the "how."

Hobart is a graduate student in a clinical psychology training program. He puts himself in Richard's shoes and asks himself, "What would I have to do to get a book written?" Here are some of his brainstormed possibilities:

- Get a book on writing and learn the basics.
- Start writing short bits on my own experience, anything that comes to mind.
- Read books written by those who conquered some kind of tragedy.
- Talk to the authors of these books.
- Find out what the pitfalls of writing are, like writer's block.
- Get a ghost writer who can translate my ideas into words.
- Write a number of very short, to-the-point pamphlets, then turn them into a book.
- Learn how to use a word-processing program both as part of my physical rehabilitation program and as a way of jotting down and playing with ideas.
- Do rough drafts of topics that interest me, and let someone else put them into shape.
- Record discussions about my own experiences with the counselor, the rehabilitation professionals, and friends and then have these transcribed for editing.
- Interview people who have turned tragedies like mine around.
- Interview professionals and the relatives and friends of people involved in personal tragedies. Record their points of view.
- Through discussion with friends get a clear idea of what this book will be about.
- Find some way of making it a bit different from similar books. What could I do that would give such a book a special slant?

1. Now do the same kind of work for each of the following cases.

Case # 1. Noma, 27, has been an intravenous drug user. She has learned that she is not only HIV positive but that she has ARC (AIDS Related Complex). She has begun taking two drugs and her symptoms are in remission. Since her family — parents, two brothers, one married, and two sisters, both married — never approved of her lifestyle, she moved to a different city. But now she wants to return to her home town and struggle with this illness where she grew up. One of her goals is reconciliation with her family. There has been very little communication with them, but she did return briefly for two of the weddings. She wants to start the work on reconciliation while she is still feeling well. She realizes that reconciliation is a two-way street and that she cannot set goals for others.

a. **Goal.** If you were Noma, what would "reconciliation with my family" look like? If accomplished, what would this goal look like? What would be in place that is not now in place? How is the goal modified by the fact that reconciliation is a two-way street? Formulate a goal that has the characteristics of a viable goal outlined earlier.

_____

_____

_____

_____

b. **Brainstorming Strategies.** What are some of the things you might do to achieve the kind of reconciliation with your family that you have outlined above? Do the brainstorming on a separate sheet of paper. Include some "wild" possibilities.

2. After you have developed your list, first share your goal with a learning partner. Give each other feedback on the quality of the goal. Note especially how you and your partner interpret "reconciliation."
3. Next share the lists of brainstormed possibilities. Working together, add several more possibilities to the combined lists. Keep your list for use in an exercise in Step III-B.
4. Discuss with your partner what you have learned from this exercise.

Case # 2. A priest was wrongfully accused of molesting a boy in his parish. Working with a counselor, he set three goals. His "now" goal was to maintain his equilibrium under stress. At the time other priests had been accused and convicted of pedophilia. He knew that in the eyes of many he would be seen to be guilty until proved innocent. With the help of the counselor and some close friends, both lay and clerical, he kept his head above water. A near-term goal was to win the case in court. He also accomplished this goal. He was acquitted, and all charges were dropped. After the trial the bishop wanted to send him to a different parish "to start fresh." He had been removed from his pastorate and was living in a different parish. But he wanted to return to the same parish and re-establish his relationship with his parishioners. After all, he had done nothing. The bishop returned him to the same parish.

a. **Goal.** If you were this man, what would "re-establishing my relationship with and reconciliation with my parishioners" look like? What would be in place that is not now in place? Formulate a goal that has the characteristics of a viable goal outlined earlier.

_____

_____

_____

_____

b. **Brainstorming Strategies.** What are some of the things you might do to re-establish your relationship with your parishioners, especially in view of the fact that some might still see you as tainted by the whole affair? Do the brainstorming on a separate sheet of paper. Include some "wild" possibilities.

2. After you have developed your list, first share your goal with a learning partner. Give each other feedback on the quality of the goal. Note especially how you and your partner might differ in your approach to re-establishing the relationship.
3. Next share the lists of brainstormed possibilities. Working together, add several more possibilities to the combined lists. Keep your list for use in an exercise in Step III-B.
4. Discuss with your partner what you have learned from this exercise.

EXERCISE 52: HELPING OTHERS BRAINSTORM STRATEGIES FOR ACTION

As a counselor, you can help your fellow trainees stimulate their imaginations to come up with creative ways of achieving their goals. In this exercise, use probes and challenges based on questions such as the following:

- **How:** How can you get where you'd like to go? How many different ways are there to accomplish what you want to accomplish?
- **Who:** Who can help you achieve your goal? What people can serve as resources for the accomplishment of this goal?
- **What:** What resources both inside yourself and outside can help you accomplish your goal?
- **Where:** What places can help you achieve your goal?
- **When:** What times or what kind of timing can help you achieve your goal? Is one time better than another?

Example: What follows are bits and pieces of a counseling session in which Angie, the counselor trainee, is helping Meredith, the client, develop strategies to accomplish one of his goals. One of Meredith's problems is that he procrastinates a great deal. He feels that he needs to manage this problem in his own life if he is to help future clients move from inertia to action. Therefore, his "good intention" is to reduce the amount of procrastination in his life. In exploring his problem, he realizes that he puts off many of the assignments he receives in class. The result is that he is overloaded at the end of the semester, experiences a great deal of stress, does many of the tasks poorly, and receives lower grades than he is capable of. While his overall goal is to reduce the total amount of procrastination in his life, his immediate goal is to be up-to-date every week in all assignments for the counseling course. He also wants to finish the major paper for the course one full week before it is due. He chooses this course as his target because he finds it the most interesting and has many incentives for doing the work on time. He presents his list of the strategies he has brainstormed on his own to Angie. After discussing this list, their further conversations sound something like this:

Angie. "You said that you waste a lot of time. Tell me more about that."
Meredith. "Well, I go to the library a lot to study, with the best intentions, but I meet friends, we kid around, and time slips away. I guess the library is not the best place to study."
Angie. "Does that suggest another strategy?"
Meredith. "Yeah, study someplace where none of my friends is around. But then I might not . . ."
Angie (interrupting): "We'll evaluate this later. Right now let's just add it to the list."

* * * * *

Angie. "I know it's your job to manage your own problems, but I assume that you could get help from others and still stay in charge of yourself. Could anyone help you achieve your goal?"
Meredith. "I've been thinking about that. I have one friend . . . we make a bit of fun of him because he makes sure he gets everything done on time. He's not the smartest one of our group, but he gets good grades because he knows how to study. I'd like to pair up with him in some way, maybe even anticipate deadlines the way he does."

* * * * *

Angie. "Your strategy list sounds a bit tame. Maybe it sounds wild to you because you're trying to change what you do."
Meredith. "I guess I could get wilder. Hmmm. I could make a contract with my counseling prof to get the written assignments in early! That would be wild for me."

Note here that Angie uses probes based on the how, who, what, when, and where probes outlined above. She also uses the "wilder possibilities" probe.

1. Divide up into groups of three: client, helper, and observer.
2. Decide in which order you will play these roles.
3. The client will briefly summarize a concern or problem and a specific goal which, if accomplished, will help him or her manage the problem or develop the unused opportunity more fully. Make sure that the client states the goal in such a way that it fulfills the criteria for a viable goal.
4. Give the client five minutes to write down as many possible ways of accomplishing the goal as he or she can think of.
5. Then help the client expand the list. Use probes and challenges based on the questions listed above.
6. Encourage the client to follow the rules of brainstorming. For instance, do not let him or her criticize the strategies as he or she brainstorms.
7. At the end of the session, stop and receive feedback from your client and the observer as to the helpfulness of your probes and challenges.
8. Switch roles and repeat the process until each has played all three roles.

Section 15
STEP III-B: BEST-FIT STRATEGIES —
WHAT ACTIONS ARE BEST FOR YOU?

The principle is simple. Strategies for action chosen from a large pool of strategies tend to be more effective than those chosen from a small pool. However, if brainstorming is successful, clients are sometimes left with more possibilities than they can handle. Therefore, once clients have been helped to brainstorm a range of strategies, they might also need help in choosing the most useful. These exercises are designed to help you help clients choose "best-fit" strategies, that is, strategies that best fit the resources, style, circumstances, and motivation level of clients. We begin with you.

EXERCISE 53: A PRELIMINARY SCAN OF BEST-FIT STRATEGIES FOR YOU

You do not necessarily need sophisticated methodologies to come up with a package of strategies that will help you accomplish a goal. In this exercise you are asked to use your common sense to make a "first cut" on the strategies you brainstormed for yourself in Exercise 50.

1. Review the strategies you brainstormed for each of the goals considered in that exercise.
2. Star the strategies that make the most sense to you. Just use common-sense judgment. Use the following example as a guideline.

Example: Let's return to the case of Ira in Exercise 50. Remember that he is the counselor trainee who wants to stop smoking. Here are the strategies he brainstormed. The ones he chooses in a preliminary common-sense scan are marked with a square (□) instead of a bullet (·).

- ☐ just stop cold turkey.
- ☐ shame myself into it, "How can I be a helper if I engage in self-destructive practices such as smoking?"
- • cut down, one less per day until zero is reached.
- • get the doctor to prescribe the new nicotine patches.
- • look at movies of people with lung cancer.
- ☐ pray for help from God to quit.
- • use those progressive filters on cigarettes.
- • switch to a brand that doesn't taste good.
- • switch to a brand that is so heavy in tars and nicotine that even I see it as too much.
- • smoke constantly until I can't stand it any more.
- • let people know that I'm quitting.
- • put an ad in the paper in which I commit myself to stopping.
- • send a dollar for each cigarette smoked to a cause I don't believe in, for instance, the "other" political party.
- • get hypnotized; through a variety of post-hypnotic suggestions have the craving for smoking lessened.
- • pair smoking with painful electric shocks.
- • take a pledge before my minister to stop smoking.
- • join a group for professionals who want to stop smoking.
- • visit the hospital and talk to people dying of lung cancer.
- • if I buy cigarettes and have one or two, throw the rest away as soon as I come to my senses.
- • hire someone to follow me around and make fun of me whenever I have a cigarette.
- • have my hands put in casts so I can't hold a cigarette.
- • don't allow myself to watch television on the days in which I have even one cigarette.
- ☐ reward myself with a week-end fishing trip once I have not smoked for two weeks.
- ☐ substitute chewing gum for smoking, starting first with nicotine-flavored gum.
- • avoid friends who smoke.
- • have a ceremony in which I ritually burn whatever cigarettes I have and commit myself to living without them.
- • suck on hard candy made with one of the new non-sugar sweeteners instead of smoking.
- ☐ give myself points each time I want to smoke a cigarette and don't; when I have saved up a number of points, reward myself with some kind of "luxury."

3. Share with a learning partner the strategies you have starred and the reasons for choosing them.

EXERCISE 54: BEST-FIT STRATEGIES: PUTTING YOURSELF IN THE CLIENT'S SHOES

In this exercise you are asked to put yourself in clients' shoes as they struggle to choose the strategies that will best enable them to accomplish their goals.

1. Review the case of Richard in Exercise 51.
2. Review the strategies that were brainstormed by Hobart, the clinical psychology trainee who put himself in Richard's shoes:

- Get a book on writing and learn the basics.
- Start writing short bits on my own experience, anything that comes to mind.
- Read books written by those who conquered some kind of tragedy.
- Talk to the authors of these books.
- Find out what the pitfalls of writing are, like writer's block.
- Get a ghost writer who can translate my ideas into words.
- Write a number of very short, to-the-point pamphlets, then turn them into a book.
- Learn how to use a word-processing program both as part of my physical rehabilitation program and as a way of jotting down and playing with ideas.
- Do rough drafts of topics that interest me, and let someone else put them into shape.
- Record discussions about my own experiences with the counselor, the rehabilitation professionals, and friends and then have these transcribed for editing.
- Interview people who have turned tragedies like mine around.
- Interview professionals and the relatives and friends of people involved in personal tragedies. Record their points of view.
- Through discussion with friends, get a clear idea of what this book will be about.
- Find some way of making it a bit different from similar books. What could I do that would give such a book a special slant?

3. If you have further strategies, add them to the list now.
4. Using your common sense, circle the strategies you believe belong in the best-fit category.
5. Jot down the reasons for your choices.
6. Share your choices and reasons with a learning partner and give each other feedback.

EXERCISE 55: USING CRITERIA TO CHOOSE BEST-FIT STRATEGIES

Just as there are criteria for crafting preferred-scenario goals and agendas (Step II-B), so there are criteria for choosing best-fit strategies. The following questions can be asked, especially when the client is having difficulty choosing from among a number of possibilities. These criteria complement rather than take the place of common sense.

- **Clarity.** Is the strategy clear?
- **Relevance.** Is it relevant to my problem situation and goal?
- **Realism.** Is it realistic? Can I do it?
- **Appeal.** Does it appeal to me?
- **Values.** Is it consistent with my values?
- **Efficacy.** Is it effective enough? Does it have bite? Will it get me there?

These questions can be recalled through the acronym CRRAVE, on the assumption that clients "crave" to accomplish their goals.

Example: Ira, the counselor trainee who wanted to quit smoking, considered the following possibility on his list: "Cut down gradually, that is, every other day eliminate one cigarette from the 30 I smoke daily. In two months, I would be free."

C - Clarity: "This strategy is very clear; I can actually see the number diminishing. It gets a 6 or 7 for clarity."

R - Relevance: "It leads inevitably to the elimination of my smoking habit, but only if I stick with it."

R - Realism: "I could probably bring this off. It would be like a game; that would keep me at it. But maybe too much like a game."

A - Appeal: "I like the idea of easing into it; but I'm quitting because I now am convinced that smoking is very dangerous. I should stop at once."

V - Values: "There is something in me that says that I should be able to quit cold turkey. That has more moral appeal to me. Gradually cutting down is for weaker people."

E - Effectiveness: "The more I draw this action program out, the more likely I am to give it up. There are too many pitfalls spread out over a two-month period."

In summary, Ira says, "I now see that only strategies related to stopping cold turkey have bite. In fact, stopping is not the hard thing. Not taking smoking up again in the face of temptation, that's the real problem." The CRRAVE criteria not only helped Ira eliminate all strategies related to a gradual reduction in smoking, but they helped him redefine his goal to "stopping and staying stopped." Sustainability is the real issue. He also noted that the brainstormed strategies he preferred referred not just to stopping but to sustainability.

1. Read the following case, put yourself in this woman's shoes, and, like Ira, use the CRRAVE criteria for determining the viability of the strategy she proposes.

Case: A young woman has been having disagreements with a male friend. Since he is not the kind of person she wants to marry, her goal is to establish a relationship with him that is less intimate, for instance, one without sexual relations. She knows that she can be friends with him but is not sure if he can be just a friend with her. She would rather not lose him as a friend. She also knows that he sees other women. She uses the CRRAVE criteria to evaluate the following strategy: "I'll call a moratorium on our relationship. I'll tell him that I don't want to see him for four months. After that we will be in a better position to re-establish a different kind of relationship, if that's what both of us want."

2. Once you have done your analysis, share your findings with a learning partner. See if the two of you can learn from your differences.
3. Now review the reasoning that the woman herself went through and her decision.

**DO NOT READ THE NEXT SECTION
BEFORE DOING STEPS 1 AND 2 OF THIS EXERCISE!**

Here, then, is her analysis:

C - Clarity: "A moratorium is quite clear; it would mean stopping all communication for four months. It would be as if one of us were in Australia for four months. But no phone calls."

R - Relevance: "Since my goal is moving into a different kind of relationship with him, stepping back to let old ties and behaviors die a bit is essential. A moratorium is not the same as ending a relationship. It leaves the door open. But it does indicate that cutting the relationship off completely could ultimately be the best course."

R - Realism: "I can stop seeing him. I think I have the assertiveness to tell him exactly what I want and stick to my decision. Obviously I don't know how realistic he will think it is. He might see it as an easy way for me to brush him off. He might get angry and tell me to forget about it."

A - **Appeal:** "The moratorium appeals to me. It will be a relief not having to manage my relationship with him for a while."

V - **Values:** "There is something unilateral about this decision, and I prefer to make decisions that affect another person collaboratively. On the other hand, I do not want to string him along, keeping his hopes for a deeper relationship and even marriage alive."

E - **Effectiveness:** "When — and if, because it depends on him, too — we start seeing each other again, it will be much easier to determine whether any kind of meaningful relationship is possible. A moratorium will help determine things one way or another." Based on her analysis, she decides to propose the moratorium to her friend.

4. Discuss her reasoning with your learning partner. In what ways does it differ from your own? If you were her helper, in what ways would you challenge her reasoning and her decision?

EXERCISE 56: USING CRITERIA TO HELP YOU CHOOSE BEST-FIT STRATEGIES

Carry out the six-step process described below for the key strategies you have starred for yourself in Exercise 53.

1. Consider the list of strategies in Exercise 53. Note the strategies you have starred.
2. Choose two starred strategies that (a) are important to you and (b) are substantial enough to merit the kind of scrutiny the CRRAVE criteria provide, that is, two key strategies.
3. Use the CRRAVE questions to check the viability of each of these strategies.
4. Share your choices and your reasoning with a learning partner. Get feedback from him or her.

EXERCISE 57: CHOOSING BEST-FIT STRATEGIES: THE BALANCE SHEET METHOD

The balance sheet is another tool you can use to evaluate different program possibilities or courses of action. It is especially useful when the problem situation is serious and you are having difficulty rating different courses of action. Review the balance sheet format in Chapter Eleven of *The Skilled Helper*.

Example: Rev. Alex M. has gone through several agonizing months re-evaluating his vocation to the ministry. He finally decides that he wants to leave the ministry and get a secular job. His decision, though painful in coming, leaves him with a great deal of peace. He now wonders just how to go about this. One possibility, now that he has made his decision, is to leave immediately. However, since this is a serious choice, he wants to look at it from all angles. He uses the decision balance sheet to evaluate the strategy of leaving his position at his present church immediately. We will not present his entire analysis (indeed, each bit of the balance sheet need not be used). Here are some of his key findings:

- **Benefits for me:** Now that I've made my decision, it will be a relief to get away. I want to get away as quickly as possible.

□ **Acceptability:** I have a right to think of my personal needs. I've spent years putting the needs of others and of the institution ahead of my own. I'm not saying that I regret this. Rather, this is now my "season," at least for a while.

□ **Unacceptability:** Leaving right away seems somewhat impulsive to me, meeting my own needs to be rid of a burden.

- **Costs for me:** I don't have a job, and I have practically no savings. I'll be in financial crisis.

  □ **Acceptability:** My frustration is so high that I'm willing to take some financial chances. Besides, I'm well educated, and the job market is good.

  □ **Unacceptability:** I will have to forgo some of the little luxuries of life for a while, but that's not really unacceptable.

- **Benefits for significant others:** The associate minister of the parish would finally be out from under the burden of these last months. I have been hard to live with. My parents will actually feel better because they know I've been pretty unhappy.

  □ **Acceptability:** My best bet is that the associate minister will be so relieved that he will not mind the extra work. Anyway, he's much better than I at getting people involved in the work of the congregation.

  □ **Unacceptability:** I can't think of any particular downside here.

- **Costs to significant social settings:** This is the hard part. Many of the things I do in this church are not part of programs that have been embedded in the structure of the church. They depend on me personally. If I leave immediately, many of these programs will falter and perhaps die because I have failed to develop leaders from among the members of the congregation. There will be no transition period. The congregation can't count on the associate minister taking over, since he and I have not worked that closely on any of the programs in question.

  □ **Acceptability:** The members of the congregation need to become more self-sufficient. They should work for what they get instead of counting so heavily on their ministers.

  □ **Unacceptability:** Since I have not worked at developing lay leaders, I feel some responsibility for doing something to see to it that the programs do not die. Some of my deeper feelings say that it isn't fair to pick up and run.

Alex goes on to use this process to help him make a decision. He finally decides to stay an extra three months and spend time with potential leaders within the congregation. He will tell them his intentions and then help them take ownership of essential programs.

1. Choose a personal goal for which you have brainstormed strategies.
2. Choose a major strategy or course of action you would like to explore much more fully. If this exercise is to be meaningful, the problem area, the goal, and the strategy or course of action in question must have a good deal of substance to them. Using the balance sheet to make a relatively inconsequential choice is a waste of time.
3. Identify the "significant others" and the "significant social settings" that would be affected by your choice.
4. Explore the possible course of action by using as much of the balance sheet as is necessary to help you make a sound decision.

Section 16
STEP III-C: MAKING PLANS —
HELPING CLIENTS DEVELOP ACTION PLANS THAT WORK

A plan is a step-by-step procedure for accomplishing each goal of an agenda. The strategies chosen in Step III-B often need to be translated into a step-by-step plan. Clients are more likely to act if they know what they are going to do first, what second, what third, and so forth. Realistic time frames for each of the steps are also essential. The plan imposes the discipline clients need to get things done. To prepare for these exercises, read Chapter Sixteen.

EXERCISE 58: SETTING UP THE MAJOR STEPS OF YOUR OWN ACTION PLAN

In this exercise you are asked to establish a plan to accomplish one of the goals you have set for yourself. First, consider the following case.

Example: Eliza, 38, a widow with two children in their upper teens, wants to get a job. In outcome terms, "job obtained and started" is her goal. However, in talking to a counselor, she soon realizes that there are a number of steps in a program leading to the accomplishment of this goal. In putting a plan together, she comes up with the following major steps of a plan.

- **Step 1: Job criteria established.** She soon discovers that she doesn't want just any kind of job. She has certain standards she would like to meet insofar as this is possible in the current job market.
- **Step 2: Resume developed.** In order to advertise herself well, she needs a high-quality resume.
- **Step 3: Job possibilities canvassed.** She needs to find out just what kinds of jobs are available that meet her general standards.
- **Step 4: A "best-possibilities" list drawn up.** She needs to draw up a list of possibilities that seem most promising in view of the job market and the standards she has worked out.
- **Step 5: Job interviews applied for and engaged in.** This includes sending out her resume. She has to find out whether she wants a particular job and whether the employer wants her.
- **Step 6: Best offer chosen and job started.** If she receives two or more offers that meet her standards, she must decide which offer to accept.
- **Contingency plan.** If the kind of job search she designs proves fruitless, she needs to know what she is to do next. She needs a backup plan.

1. Review the work you did for your own constructive change program in Exercises 50 and 53.
2. Once more, indicate the goal you want to accomplish.

---

3. List below the best-fit strategies you chose for yourself in Exercise 53.

_____

_____

_____

_____

_____

_____

4. Take these strategies and turn them into a workable plan. What are you going to do first, second, third, and so forth? Put some timelines on your plan. When are you going to accomplish each step?

_____

_____

_____

_____

_____

_____

_____

_____

_____

_____

_____

_____

5. Share your plan with a learning partner. Give each other feedback on the quality of the plan. Use probes in talking to each other to help the other discover the most difficult parts of the plan. What is the probability of overall success? How much would you bet on the full implementation of the plan?

6. In light of the feedback and discussion, what changes would you make in the plan?

_____

_____

_____

_____

_____

EXERCISE 59: SETTING UP THE MAJOR STEPS OF AN ACTION PLAN

Let us return to the case of Richard, the man who wanted to write a book about hope as part of his overall rehabilitation plan after an accident. As you recall, Hobart, putting himself in Richard's shoes, brainstormed the following strategies.

- Get a book on writing and learn the basics.
- Start writing short bits on my own experience, anything that comes to mind.
- Read books written by those who conquered some kind of tragedy.
- Talk to the authors of these books.
- Find out what the pitfalls of writing are, like writer's block.
- Get a ghost writer who can translate my ideas into words.
- Write a number of very short, to-the-point pamphlets, then turn them into a book.
- Learn how to use a word-processing program both as part of my physical rehabilitation program and as a way of jotting down and playing with ideas.
- Do rough drafts of topics that interest me, and let someone else put them into shape.
- Record discussions about my own experiences with the counselor the rehabilitation professionals, and friends and then have these transcribed for editing.
- Interview people who have turned tragedies like mine around.
- Interview professionals and the relatives and friends of people involved in personal tragedies. Record their points of view.
- Through discussion with friends get a clear idea of what this book will be about.
- Find some way of making it a bit different from similar books. What could I do that would give such a book a special slant?

1. Place yourself in Richard's shoes.
2. Add further strategies as you see fit.
3. Identify key best-fit strategies.

_____

_____

_____

_____

_____

_____

_____

4. Turn these into the major steps of an overall plan. What are the major steps in an overall plan? What needs to be done first, second, third, and so forth? Put some timelines on the steps of the plan.

_____

_____

_____

_____

_____

_____

_____

_____

_____

_____

5. Share your finished product with a learning partner. Give each other feedback on the quality of the plan. Pool the best features of both plans. Then, on separate paper, write out the new plan.

EXERCISE 60: SUBGOALS: DIVIDE AND CONQUER

If a goal is complex, for instance, changing careers or doing something about a deteriorating marriage, the plan to achieve it will have a number of major steps. In this case a divide-and-conquer strategy is useful. That is, the complex goal (the improvement of the marriage) is divided up into a number of subgoals. In a marriage, a more equitable division of household chores might be such a subgoal. It is one goal in the total "package" of goals that will constitute the improved marriage.

In this exercise you are asked to pretend that, like Eliza, the widow in Exercise 58, you are searching for a job.

1. In order to find a job, you must draw up a resume. What is the step-by-step process you would engage in to develop a resume?

_____

_____

_____

_____

_____

2. In order to find a job, you would also have to canvass for open positions. What is the step-by-step process you would use to canvass for jobs?

_____

_____

_____

_____

_____

_____

_____

_____

3. Share your plan for each of these with a learning partner. Pool your approaches and come up with a better plan for each subgoal. Note, however, that you may have different needs. Your plans should not be exactly the same.

EXERCISE 61: FORMULATING PLANS FOR THE MAJOR STEPS OF A COMPLEX GOAL

In this exercise you are asked to spell out the action steps for two of the subgoals leading up to the accomplishment of some complex goal of your own.

Example: Lynette, a clinical psychology student in a counselor training program, has discovered that she comes across as quite manipulative both to her instructors and to her classmates. She believes that she has developed the style as a response to the less than enthusiastic reception she received whenever she encroached on "male" territory, whether at home, school, or in the workplace. However, she realizes that this style is contrary to the values she wants to permeate helper-client relations. She would end up doing to clients what she resents others doing to her. One of the goals she sets in order to manage her manipulative ways is to establish a collaborative style in dealing with teachers, classmates, and clients. This would be a major step in changing her overall manipulative style. The essence of collaboration, as she sees it, is mutual understanding of the issues and mutual decision making. Her current style in working with others is to make the decisions herself while letting the other party think that he or she is having a say. Here are the steps in her plan to implement this style change:

- Get a very clear picture of the difference between my current style and my preferred style.
- Learn a decision-making process based on mutuality (such as the negotiation process offered by Fisher and Ury in their popular book *Getting to Yes*).
- Practice mutual decision making in non-critical situations, that is, in situations in which I do not feel the other has any intention of taking advantage of me.
- Develop a list of the decision-making situations in which I must change my style.
- Plan out ahead of time what I need to do to change my interaction style in each critical encounter, for instance, in discussing a major project with an instructor in the counseling program.
- Review with myself how well I actually used the new style.
- Get feedback from the other party from time to time to see whether the process is experienced by others as mutual.

1. Evaluate the steps she has laid out. If she were to ask you for your help, what feedback would you give her about her plan? What changes would you make?
2. Now do the same for two of the major steps of a plan you have developed to achieve some complex goal related to becoming a more effective helper.

a. A key goal of mine.

---

b. One major step in the plan to accomplish this goal.

_____

_____

_____

_____

c. A brief description of the steps I would take to achieve this subgoal.

_____

_____

_____

_____

_____

_____

_____

d. A second major step in the plan to accomplish this goal.

_____

_____

_____

_____

e. A brief description of the steps I would take to achieve this subgoal.

_____

_____

_____

_____

_____

_____

_____

_____

EXERCISE 62: DEVELOPING THE RESOURCES TO IMPLEMENT YOUR PLANS

Plans can be venturesome, but they must also be realistic. Most plans call for resources of one kind or another. In this exercise you are asked to review some of the goals you have established for yourself in previous exercises and the plans you have been formulating to implement these goals with a view to asking yourself, "What kind of resources do I need to develop to implement these plans?" For instance, you may lack the kinds of skills needed to implement a program. If this is the case, the required skills constitute the resources you need.

1. **Indicate a goal** you would like to implement in order to manage some concern or problem situation in some way. Consider the following example. Mark is trying to manage his physical well-being better. He has headaches that disrupt his life. "Frequency of headaches reduced" is one of his goals, a major step toward getting into better physical shape. "The severity of headaches reduced" is another.

2. **Outline a plan** to achieve this goal. The plan may call for resources you may not have or may not have as fully as you would like. Consider Mark once more. Relaxing both physically and psychologically at times of stress and especially when he feels the "aura" that indicates a headache is on its way is one strategy for achieving his goal. Discovering and using the latest drugs to help him control his kind of headache is another part of his plan.

3. **Indicate the resources** you need to develop to implement the plan. For instance, Mark needs the skills associated with relaxing. Furthermore, since he allows himself to become the victim of stressful thoughts, he also needs some kind of thought-control skills. He does not possess either set of skills. Finally, he needs a doctor with whom he can discuss the kind of headaches he gets and possible drugs available for helping control them.

4. **Summarize a plan** that would enable you to develop some of these resources. In one program offered through the school, Mark learns the skills of systematic relaxation and skills related to controlling self-defeating thoughts. The Center for Student Services refers him to a doctor who specializes in headaches.

Problem situation #1

a. Your problem-managing or opportunity-developing goal.

_____

_____

_____

_____

b. Skills or other resources you need to accomplish this goal.

_____

_____

_____

_____

c. Summarize a plan that can help you develop or get the resources you need.

_____

_____

_____

_____

_____

Problem situation #2

a. Your problem-managing or opportunity-developing goal.

_____

_____

_____

_____

b. Skills or other resources you need to accomplish this goal.

_____

_____

_____

_____

c. Summarize a plan that can help you develop or get the resources you need.

_____

_____

_____

_____

_____

_____

EXERCISE 63: HELPING OTHERS DEVELOP THE RESOURCES NEEDED FOR IMPLEMENTING PLANS

In this exercise you are asked to help clients develop the kinds of resources they need to pursue their principal goals.

1. Mildred and Tom are having trouble with their marriage. They do not handle decisions about finances and about sexual behavior well. Fights dealing with these two areas are frequent. They both agree that their marriage would be better without these fights, and they realize that collaborative decision making with respect to sex and finances would be an ideal. But they have never communicated very well with each other.

a. Given the brief picture outlined above, what kind of resources do Mildred and Tom need to improve their marriage?

_____

_____

_____

_____

_____

b. Summarize a program that will help them develop one or two of these resources.

_____

_____

_____

_____

_____

2. Todd feels bad about his impoverished social life. He is now in his late twenties and has no intimate female friend and no close friends of either sex. He feels lonely a great deal of the time. Some of the goals he set for himself involve joining social groups, developing wider circles of acquaintances, and establishing some close friendships.

a. What kind of resources does he need for effective group participation?

_____

_____

_____

_____

_____

b. Summarize a plan that will help him develop these resources.

_____

_____

_____

_____

c. What resources does he need to establish and maintain closer and even intimate relationships?

_____

_____

_____

d. Summarize a plan that might help him develop these resources.

_____

_____

_____

_____

_____

Section 17
MAKING THINGS WORK —
HELPING CLIENTS GET WHAT THEY NEED AND WANT

There is a huge difference between talking about action and action itself. Some clients develop challenging goals, excellent strategies, and elegant plans and then stop short of using them effectively to get to their final destinations — valued outcomes, that is, outcomes that make a difference in their lives. Once a workable plan has been developed to accomplish a goal or a subgoal, clients must act. They must implement the plan.

Clients as Tacticians. There are a number of things you can do to help clients act. You can help them become effective tacticians in their everyday lives. In the exercises in this section, tactics, that is, the art or skill of employing available means to accomplish a goal *in the face of changing circumstances*, is the focus. Tactics is a military term. Not a bad choice, perhaps, because the work of implementing strategies and plans often enough resembles combat. Clients are more likely to implement plans if they can adapt them to changing conditions.

Sustained Action. When clients are trying to manage more complex problem situations, such as a deteriorating marriage or change in lifestyle, then the issue of **sustained** action is important. They must not only begin to act but continue to act even in the face of temptations to slack off. The exercises in Section 17 will help you prepare clients to grapple with temptations to quit and to sustain problem-managing and opportunity-developing action on their own.

EXERCISE 64: LEARNING FROM FAILURES

As suggested in the text, inertia and entropy dog all of us in our attempts to manage problem situations and develop unexploited opportunities. There is probably no human being who has not failed to carry through on some self-change project. This exercise assumes two things, that we have failed in some self-change project and that we can learn from our failures.

Example: Miguel kept saying that he wanted to leave his father's business and strike out on his own, especially since he and his father had heated arguments over how the business should be run. He earned an MBA in night school and talked about becoming a consultant to small, family-run businesses. A medium-sized consulting firm offered him a job. He accepted on the condition that he could finish up some work in the family business. But he always found "one more" project in the family business that needed his attention. All of this came out as part of his story, even though his main concern was the fact that his woman friend of five years had given him an ultimatum: marriage or forget about the relationship.

Finally, with the help of a counselor, Miguel makes two decisions: to take the job with the consulting firm and to agree to break off the relationship with his woman friend because he is still not seriously entertaining marriage as an immediate possibility.

However, in the ensuing year Miguel never gets around to taking the new job. He keeps finding tasks to do in his father's company and keeps up his running battle with his father. Obviously both of them, father and son, were getting something out of this in some twisted way. As to his relationship with his woman friend, the two of them broke it off four different times during that year until she finally left him and got involved with another man.

Some of Miguel's learnings. Since Miguel and his counselor were not getting anywhere, they decided to break off their relationship for a while. However, when his woman friend definitively broke off their relationship, Miguel was in such pain that he asked to see the counselor again. The first thing the counselor did was to ask Miguel what he had learned from all that had happened, on the assumption that these learnings could form the basis of further efforts. Here are some of his learnings:

- I hate making decisions that have serious action implications.
- I pass myself off as an adventuresome, action-oriented person, but at root I prefer the status quo.
- I am very ambiguous about facing the developmental tasks of an adult my age. I have liked living the life of a 17-year-old at age 30.
- I enjoy the stimulation of the counseling sessions. I enjoy reviewing my life with another person and developing insights, but this process involves no real commitment to action on my part.
- Not taking charge of my life and acting on goals has led to the pain I am now experiencing. In putting off the little painful actions that would have served the process of gradual growth I have ended up in great pain. And it could all happen again.

Notice that Miguel's learnings are about himself, his problem situations, and his way of participating in the helping process.

1. Recall some significant self-change project over the last few years that you abandoned in one way or another.
2. Picture as clearly as possible the forces at work that led to the project's ending in failure, if not with a bang, then with a whimper.
3. In reviewing your failed efforts, jot down what you learned about yourself and the process of change.
4. Share your learnings with a learning partner. Help each other (a) discover further lessons in the review of the failed project, and (b) discover what could have been done to keep the project going.

122

a. **The self-change project that failed.**

_____

_____

_____

_____

b. **Principal reasons for failure.**

_____

_____

_____

c. **What you learned about yourself as an agent of change in your own life.**

_____

_____

_____

_____

d. **What could have been done to keep the project going?**

_____

_____

_____

_____

EXERCISE 65: IDENTIFYING AND COPING WITH OBSTACLES TO ACTION

As suggested in an earlier exercise, "forewarned is forearmed" in the implementation of any plan.

1. Picture yourself trying to implement some action strategy or plan in order to accomplish a problem-managing goal. As in the example below, jot down what you actually see happening.

2. As you tell the story, describe the pitfalls or snags you see yourself encountering along the way. Some pitfalls involve inertia, that is, not starting some step of your plan; others involve entropy, that is, allowing the plan to fall apart over time.

3. Design some strategy to handle any significant snag or pitfall you identify.

Example: Justin has a supervisor at work who, he feels, does not like him. He says that she gives him the worst jobs, asks him to put in overtime when he would rather go home, and talks to him in demeaning ways. In the problem exploration phase of counseling, he discovered that he probably reinforces her behavior by buckling under, by giving signs that he feels hurt but helpless, and by failing to challenge her in any direct way. He feels so miserable at work that he wants to do something about it. One option is to move to a different department, but to do so he must have the recommendation of his immediate supervisor. Another possibility is to quit and get a job elsewhere, but he likes the company and that would be a drastic option. A third possibility is to deal with his supervisor more directly. He sets goals related to this third option.

One major step in working out this overall problem situation is to seek out an interview with his supervisor and tell her, in a strong but nonpunitive way, his side of the story and how he feels about it. Whatever the outcome, his version of the story would be on record. The counselor asks him to imagine himself doing all of this. What snags does he run into? Some of the things he says are:

- "I see myself about to ask her for an appointment. I see myself hesitating to do so because she might answer me in a sarcastic way. Also, others are usually around, and she might embarrass me and they will want to know what's going on, why I want to see her, and all that. I tell myself that I had better wait for a better time to ask."

- "I see myself sitting in her office. Instead of being firm and straightforward, I'm tongue-tied and apologetic. I forget some of the key points I want to make. I let her brush off some of my complaints and in general let her control the interaction."

a. How can he prepare himself to handle the obstacles or snags he sees in his first statement? Then what could he do in the situation itself?

_____

_____

_____

_____

_____

b. How can he prepare himself to handle the pitfalls mentioned in his second statement? What could he do in the situation itself?

_____

_____

_____

_____

_____

Personal Situation # 1.

a. Consider some plan or part of a plan you want to implement. In your mind's eye see yourself moving through the steps of the plan. What obstacles or snags do you encounter? Jot them down.

_____

_____

_____

_____

_____

b. Indicate how you might prepare yourself to handle a significant obstacle or pitfall and what you might do in the situation itself to handle it.

_____

_____

_____

_____

_____

Personal Situation # 2.

a. Consider some plan or part of a plan you want to implement. In your mind's eye see yourself moving through the steps of the plan. What obstacles or snags do you encounter? Jot them down.

_____

_____

_____

_____

_____

b. Indicate how you might prepare yourself to handle a significant obstacle or pitfall and what you might do in the situation itself to handle it.

_____

_____

_____

_____

_____

EXERCISE 66: FORCE-FIELD ANALYSIS AT THE SERVICE OF ACTION

In this exercise you are asked to identify forces "in the field," that is, out there in clients day-to-day lives, that might help them implement strategies and plans, together with forces that might hinder them. The former are called "facilitating forces" and the latter "restraining forces." The use of force-field analysis to prepare for action is an application of the adage "forewarned is forearmed." Read the section on force-field analysis in Chapter Seventeen of *The Skilled Helper*. Then review the following case.

Example: Ira, as we have seen earlier, wants to stop smoking. He has also expanded his goal from merely "stopping" to "staying stopped." He has formulated a step-by-step plan for doing so. Before taking the first step, he uses force-field analysis to identify facilitating and restraining forces in his everyday life.

Some of the facilitating forces identified by Ira:
- my own pride.
- the satisfaction of knowing I'm keeping a promise I've made to myself.
- the excitement of a new program, the very "newness" of it.
- the support and encouragement of my wife and my children.
- the support of two close friends who are also quitting.
- the good feeling of having that "gunk" out of my system.
- the money saved and put aside for more reasonable pleasures.
- the ability to jog without feeling I'm going to die.

Some of the restraining forces identified by Ira:
- the craving to smoke that I take with me everywhere.
- seeing other people smoke.
- danger times: when I get nervous, after meals, when I feel depressed and discouraged, when I sit and read the paper, when I have a cup of coffee, at night watching television.

- being offered cigarettes by friends.
- when the novelty of the program wears off (and that could be fairly soon).
- increased appetite for food and the possibility of putting on weight.
- my tendency to rationalize and offer great excuses for my failures.
- the fact that I've tried to stop smoking several times before and have never succeeded.

1. Review a goal or subgoal and the plan you have formulated to accomplish it.
2. Picture yourself "in the field" actually trying to implement the steps of the plan.
3. Identify the principal forces that are helping you reach your goal or subgoal.
4. Identify the principal forces that are hindering you from reaching your goal or subgoal.

Personal Situation # 1.

a. Spell out a goal or subgoal you want to accomplish.

_____

_____

_____

b. Picture yourself in the process of implementing the plan formulated to achieve the goal. List the facilitating forces that could help you to carry out the plan.

_____

_____

_____

_____

_____

c. List the restraining forces that might keep you from carrying out the plan.

_____

_____

_____

_____

_____

Personal Situation # 2.

a. Spell out a goal or subgoal you want to accomplish.

_____

_____

_____

b. Picture yourself in the process of implementing the plan formulated to achieve the goal. List the facilitating forces that could help you to carry out the plan.

_____

_____

_____

_____

c. List the restraining forces that might keep you from carrying out the plan.

_____

_____

_____

_____

5. Finally, share your findings with a learning partner. Use empathy, probing, and challenge to help each other clarify these two sets of forces.

EXERCISE 67: BOLSTERING FACILITATING FORCES

Once you have identified the principal facilitating and restraining forces, you can determine how to bolster critical facilitating forces and neutralize critical restraining forces. In this exercise you are asked to devise ways of bolstering critical facilitating forces.

Example: Klaus is an alcoholic who wants to stop drinking. He joins Alcoholics Anonymous. During a meeting he is given the names and telephone numbers of two people whom he is told he may call at any time of the day or night if he feels he needs help. He sees this as a critical facilitating force — just knowing that help is around the corner when he needs it. He wants to strengthen this facilitating force.

- First of all, he sees being able to get help anytime as a kind of dependency, and so he talks out the negative feelings he has about being dependent in this way with a counselor. In talking, he soon realizes that it is a temporary form of dependency that is instrumental in achieving an important goal, developing a pattern of sobriety.
- Second, he calls the telephone numbers a couple of times when he is not in trouble just to get the feel of doing so.
- Third, he puts the numbers in his wallet, he memorizes them, and he puts them on a piece of paper and carries them in a medical bracelet that tells people who might find him drunk that he is an alcoholic trying to overcome his problem.
- Finally, he calls the telephone numbers a couple of times when the craving for alcohol is high and his spirits are low. That is, he gets used to this as a temporary resource.

1. Review your list of facilitating forces from Personal Situation # 1 or # 2 in Exercise 66 that you see as capable of making a difference in the implementation of a program and that you believe you can strengthen in some way.
2. Indicate what you could do to strengthen one or more critical facilitating force.
3. Share what you intend to do with a learning partner. Using empathy, probes, and challenges, help each other strengthen the plan.

a. Briefly describe one or two key facilitating forces from Personal Situation #1 in Exercise 66 that you would like to strengthen.

_____

_____

_____

b. Indicate how you would like to go about strengthening these key facilitating forces.

_____

_____

_____

_____

EXERCISE 68: MANAGING KEY RESTRAINING FORCES

Sometimes it is helpful to try to neutralize or reduce the strength of critical restraining forces. Act as a consultant to the woman in the example.

Example: Ingrid is on welfare, but she has a goal of getting a job. Part of her plan is to apply for and go to job interviews. However, she ends up missing a number of the interviews. By examining her behavior, she learns that there are at least two critical restraining forces. One is that she has a poor self-image: she thinks she looks ugly and that the interviewer won't give her a fair chance simply because of her looks. Another is that at the last moment she thinks of a number of "important" tasks that must be done, for instance, visiting her ailing mother, before she can do anything else. She does these tasks instead of going to the interview.

a. How might Ingrid handle the problem of feeling ashamed of her looks?

_____

_____

_____

_____

b. How might Ingrid handle the problem of putting "important" tasks ahead of going to job interviews?

_____

_____

_____

_____

1. On separate paper, identify some key restraining forces you face in the pursuit of important personal goals and indicate what you can do to minimize these forces.
2. Share your findings with a learning partner and help each other to find even better ways of anticipating and dealing with substantial restraining forces.

EXERCISE 69: USING SUPPORTIVE AND CHALLENGING RELATIONSHIPS

Key people in the day-to-day lives of clients can play an important part in helping them stay on track as they move toward their goals. If part of a client's problem is that he or she is "out of community," then a parallel part of the helping process should be to help the client develop supportive human resources in his or her everyday life. In this exercise you are asked to look at strategies and plans from the viewpoint of these human resources. People can provide both support and challenge.

Example: Enid, a 40-year-old single woman, is making a great deal of progress in terms of controlling her drinking through her involvement with an AA program. But she is also trying to decide what she wants to do about a troubled relationship with a man. In fact, her drinking was, in part, an ineffective way of avoiding the problems in the relationship. She knows that she no longer wants to tolerate the psychological abuse she has been getting from him, but she also fears the vacuum she will create by cutting the relationship off. She is, therefore, trying to develop some possibilities for a better relationship. She also realizes that ending the relationship might be the best option. Because of counseling, she has been much more assertive in the relationship. She now cuts off contact whenever he becomes abusive. That is, she is already engaging in a series of "little actions" that help her better manage her life and discover further possibilities. Finally, since this is not the only time she has experienced psychological abuse, she is beginning to wonder what it is about herself that in some way almost draws contempt. She has a sneaking feeling that the contempt of others might merely mirror the way she feels about herself. She has also begun to wonder why she has stayed in a safe but low-paying job so long. She realizes that at work she is simply taken for granted. In other words, there are some issues that she has yet to explore.

In summary, because Enid has a troubled relationship with herself, she is likely to have troubled relationships with others, in this case her male friend. At time she resorts to drinking to "manage" her relationships both to herself and to others. It is an ineffective management technique, providing temporary relief from pain but, in the long run, causing even more pain.

Over the course of two years, the counselor helps Enid develop human resources for both support and challenge.

- She moves from one-to-one counseling to group counseling with occasional one-to-one sessions. Group members provide a great deal of both support and challenge.
- She begins attending church. In the church she attends a group something like Alcoholics Anonymous.
- Through one of the church groups she meets and develops a friendship with a 50-year-old woman who has "seen a lot of life" herself. She challenged Enid when she began feeling sorry for herself. She also introduced Enid to the world of art.
- She does some volunteer work at an AIDS center. The work challenges her, and there is a great deal of camaraderie among the volunteers.

Since one of Enid's problems is that she is "out of community," these human resources constitute part of the "solution." Rather, all these contacts give her multiple opportunities to get back into community and both give and get support and challenge.

1. Summarize some goal you are pursuing and the action plan you have developed to get you there.
2. Identify the human resources that are already part of that plan.
3. Indicate the ways in which people provide support for you as you implement your plan.
4. Indicate ways in which people challenge you to keep to your plan or even change it when appropriate.
5. What further support and challenge would help you stick to your program?
6. Indicate ways of tapping into or developing the people resources needed to provide that support and challenge.

a. Summarize your goal and action plan.

_____

_____

_____

_____

b. Identify the human resources that are already part of that plan.

_____

_____

_____

c. Indicate the ways in which people might provide further support and challenge for you as you implement your plan.

_____

_____

_____

_____

_____

APPENDIX ONE

EXERCISES IN SELF-ASSESSMENT

The exercises in this section are designed to help you tell your own story, that is, to help you identify the issues, problems, and concerns of your own life that might stand in the way of helping others. Thoughtful execution of some of the exercises in this section will give you a list of issues, neither too superficial nor too intimate, to discuss and work on in the training group. The training program is a golden opportunity for you to grow and develop.

Focusing on the issues that might stand in the way of being a skilled helper with your clients is in keeping with the values of competence, pragmatism, genuineness, and self-responsibility discussed in Chapter Three of the text. Later on, some of these exercises can be used with clients to help them identify and clarify their concerns.

THE KINDS OF CONCERNS TRAINEES DISCUSS

Here, in statement form, are some of the kinds of problems, issues, and concerns that trainees have dealt with during training programs.

- I'm shy. My shyness takes the form of being afraid to meet strangers and being afraid to reveal myself to others.
- I'm a fairly compliant person. Others can push me around and get away with it.
- I get angry fairly easily and let my anger spill out on others in irresponsible ways. I think my anger is often linked to not getting my own way.
- I'm a lazy person. I find it especially difficult to expend the kind of energy necessary to listen to and get involved with others.
- I'm somewhat fearful of persons of the opposite sex. This is especially true if I think they are putting some kind of demand on me for closeness. I get nervous and try to get away.
- I'm a rather insensitive person, or so I have been told. I'm a kind of bull-in-the-china-shop type. Not much tact.

- I'm overly controlled. I don't let my emotions show very much. Sometimes I don't even want to know what I'm feeling myself.
- I like to control others, but I like to do so in subtle ways. I want to stay in charge of interpersonal relationships at all times.
- I have a strong need to be liked by others. I seldom do anything that might offend others or that others would not approve of. I have a very strong need to be accepted.
- I have few positive feelings about myself. I put myself down in a variety of ways. I get depressed a lot.
- I never stop to examine my values. I think I hold some conflicting values. I'm not even sure why I'm interested in becoming a helper.
- I feel almost compelled to help others. It's part of my religious background. It's as if I didn't even have a choice.
- I'm sensitive and easily hurt. I think I send out messages to others that say "be careful of me."
- I'm overly dependent on others. My self-image depends too much on what others think of me.
- A number of people see me as a "difficult" person. I'm highly individualistic. I'm ready to fight if anyone imposes on my freedom.
- I'm anxious a lot of the time. I'm not even sure why. My palms sweat a lot in interpersonal situations.
- I see myself as a rather colorless, uninteresting person. I'm bored with myself at times, and I assume that others are bored with me.
- I'm somewhat irresponsible. I take too many risks, especially risks that involve others. I'm very impulsive. That's probably a nice way of saying that I lack self-control.
- I'm very stubborn. I have fairly strong opinions. I argue a lot and try to get others to see things my way. I argue about very little things.
- I don't examine myself or my behavior very much. I'm usually content with the way things are. I don't expect too much of myself or of others.
- I can be sneaky in my relationships with others. I seduce people in different ways—not necessarily sexual—by my "charm." I get them to do what I want.
- I like the good life. I'm pretty materialistic, and I like my own comfort. I don't often go out of my way to meet the needs of others.
- I'm somewhat lonely. I don't think others like me, if they think about me at all. I spend time feeling sorry for myself.
- I'm awkward in social situations. I don't do the right thing at the right time. I don't know what others are feeling when I'm with them and I guess I seem callous.
- Others see me as "out of it" a great deal of the time. I guess I am fairly naive. Others seem to have deeper or more interesting experiences than I do. I think I've grown up too sheltered.
- I'm stingy with both money and time. I don't want to share what I have with others. I'm pretty selfish.
- I'm somewhat of a coward. I sometimes find it hard to stand up for my convictions even when I meet light opposition. It's easy to get me to retreat.
- I hate conflict. I'm more or less a peace-at-any-price person. I run when things get heated up.
- I don't like it when others tell me I'm doing something wrong. I usually feel attacked, and I attack back.

This list is not exhaustive, but you can use it to stimulate your thinking about yourself and the kinds of dissatisfactions, problems, or concerns you may have about yourself, especially concerns that might relate to your effectiveness as a helper. Exercises A through G will help you assess your satisfactions and dissatisfactions with yourself and your behavior. You can then choose the issues that you would like to explore during the training sessions.

STRENGTHS AND SOFT SPOTS

All of us have strengths and soft spots, areas in which we could use some improvement. I might well be a decisive person and that is a strength for me as a helper, but, in being decisive, I might push others too hard and that is a soft spot needing improvement. In this case, a strength pushed too hard becomes a soft spot.

EXERCISE A: STRENGTHS AND SOFT SPOTS IN MY LIFE

Sometimes a very simple structure can help you and your clients identify the major dimensions of a problem situation. This exercise asks you to identify some of the things that are not going as well as you would like them to go in your life and some of the things you believe you are handling well. It is important right from the beginning to help clients become aware of their resources and successes as well as their problems and failures. Problems can be handled more easily if they are seen in the wider context of resources and successes.

In this exercise, merely jot down, in whatever way they come to you, things that are going right and things that could be going better for you. In order to stress the positive, see if you can write down at least two things that are going right for every thing that needs improvement. Read the list in the example that follows and then do your own. Don't worry whether the problems or concerns you list are really important. Jot down whatever comes to your mind. Your own list may include items similar to those in the example, but it may be quite different because it will reflect you and not someone else.

Example

Strengths	Soft Spots
I have a lot of friends.	I seem to have a very negative attitude toward myself.
I have a decent job and people like the work I do.	I get dependent on others much too easily.
Others can count on me; I'm dependable.	My life seems boring too much of the time.
I have a reasonable amount of intelligence.	I am afraid to take risks.
I have no major financial difficulties; I'm secure.	
My wife and I get along fairly well.	
I am very healthy.	
My belief in God gives me a kind of center in life, a stability.	

Strengths	Soft Spots
_____	_____
_____	_____
_____	_____
_____	_____
_____	_____
_____	_____
_____	_____
_____	_____
_____	_____
_____	_____

DEVELOPMENTAL TASKS

Many of the clients you see will be struggling with developmental concerns like those you have struggled with or are currently working on. Like you, they will have both strengths and "soft spots" in coping with the developmental tasks of life.

EXERCISE B: REVIEWING SOME BASIC DEVELOPMENTAL TASKS

In this exercise you are asked to consider your experience with ten major developmental tasks of life. First reflect on your experience in these developmental areas and then apply what you have learned to your role as a helper of others. Use extra paper as needed.

1. **COMPETENCE. What do I do well?** Do I see myself as a person who is capable of getting things done? Do I have the resources needed to accomplish goals I set for myself? In what areas of life do I excel? In what areas of life would I want to be more competent than I am?

Strengths	**Soft Spots**
_____	_____
_____	_____
_____	_____
_____	_____

2. **AUTONOMY: Can I make it on my own?** Can I get things done on my own? Do I avoid being overly dependent or independent? Am I reasonably interdependent in my work and social life? When I need help, do I find it easy to ask for it? In what social settings do I find myself most dependent? counterdependent? independent? interdependent?

Strengths	**Soft Spots**
_____	_____
_____	_____
_____	_____
_____	_____

3. **VALUES: What do I believe in?** What are my principal values? Do I allow for reasonable changes in my value system? Do I put my values into practice? Do any of the values I hold conflict with others? In what social settings do I pursue the values that are most important to me?

Strengths	**Soft Spots**
_____	_____
_____	_____
_____	_____
_____	_____

4. **IDENTITY: Who am I in this world?** Do I have a good sense of who I am and the direction I'm going in life? Do the ways that others see me fit with the ways in which I see myself? Do I have some kind of center that gives meaning to my life? In what social settings do I have my best feelings for who I am? In what social settings do I lose my identity? In what ways am I confused or dissatisfied with who I am?

Strengths	Soft Spots
_____	_____
_____	_____
_____	_____
_____	_____

5. **INTIMACY. What are my closer relationships like?** What kinds of closeness do I have with others? To what extent are there degrees of closeness in my life—acquaintances, friends, and intimates? What is my life in my peer group like? How well do I get along with others? What concerns do I have about my interpersonal life?

Strengths	Soft Spots
_____	_____
_____	_____
_____	_____
_____	_____

6. **SEXUALITY. Who am I as a sexual person?** To what degree am I satisfied with my sexual identity, my sexual preferences, and my sexual behavior? How do I handle my sexual needs and wants? What social settings influence the ways I act sexually?

Strengths	Soft Spots
_____	_____
_____	_____
_____	_____
_____	_____

7. **LOVE, MARRIAGE, FAMILY. What are my deeper commitments like?** What is my marriage like? How do I relate to family and relatives? How do I feel about the quality of my family life? If not married, in what ways do I look forward to marriage? What misgivings do I have?

Strengths	Soft Spots
_____	_____
_____	_____
_____	_____
_____	_____

8. CAREER. What is the place of work in my life? How do I feel about the way I am preparing myself for a career or the career I am currently pursuing? What do I get out of work? What am I like in the workplace? How does it affect me? What impact do I have there?

Strengths	Soft Spots
_____	_____
_____	_____
_____	_____

9. INVESTMENT IN THE WIDER COMMUNITY. How big is my world? How do I invest myself in the world outside of friends, work, and the family? What is my neighborhood like? Do I have community, civic, political, social involvements or concerns? In what ways am I optimistic about the world? In what ways am I cynical?

Strengths	Soft Spots
_____	_____
_____	_____
_____	_____

10. LEISURE. What do I do with my free time? Do I feel that I have sufficient free time? How do I use my leisure? What do I get out of it? In what social settings do I spend my free time?

Strengths	Soft Spots
_____	_____

_____   _____

_____   _____

_____   _____

Many of the clients you see will be struggling with similar developmental concerns. Like you, they will have both strengths and "soft spots" in these areas.

In your opinion, which of the strengths you have noted will help you be a more effective counselor? In what specific ways?

_____

_____

_____

_____

_____

_____

_____

_____

_____

_____

_____

In your opinion, which of the soft spots you have noted might stand in the way of your being an effective helper? In what specific ways?

_____

_____

_____

_____

THE SOCIAL SETTINGS OF LIFE

We belong to and participate in a number of different social settings in life: family, circle of friends, clubs, church groups, school groups, and the like. We are also affected by what goes on in our neighborhoods and the cities and towns in which we live. Larger systems, such as state and national governments, have their ways of entering our lives. We manage some of our interactions in these settings quite well; we have soft spots in others.

EXERCISE C: CONFLICTS IN THE SOCIAL SETTINGS OF LIFE

This is another exercise that will help you review areas in which clients have problems and also help you identify strengths and soft spots that relate to your role as helper.

1. **Chart the social settings of life.** Since you are a member of a number of different social settings and since each places certain demands on you, conflicts can arise between settings. In this exercise you are asked to write your name in the middle of a sheet of paper. Then, as in the example (Figure 1), draw spokes out to the various social settings of your life. The person in the example is Mitch, 45, a principal of an inner-city high school in a large city. He is married and has two teenage sons. Neither attends the high school of which he is principal. He is seeing a counselor because of exhaustion and bouts of hostility and depression. He has had a complete physical check-up, and there is no evidence of any medical problem.

2. **Review expectations, demands, concerns.** Now take each social setting and write down the expectations people have of you in that setting, the demands they place on you, the concerns you have, the dissatisfactions expressed to you. For instance, some of things Mitch writes are:

Faculty
- Some faculty members want a personal relationship with me, and I have neither the time nor the desire.
- Some faculty members have retired on the job. I don't know what to do with them.
- Some of the white faculty members are suspicious of me and distant just because I'm black.
- One faculty member wrote the district superintendent and said that I was undermining her reputation with other faculty members. This is not true.

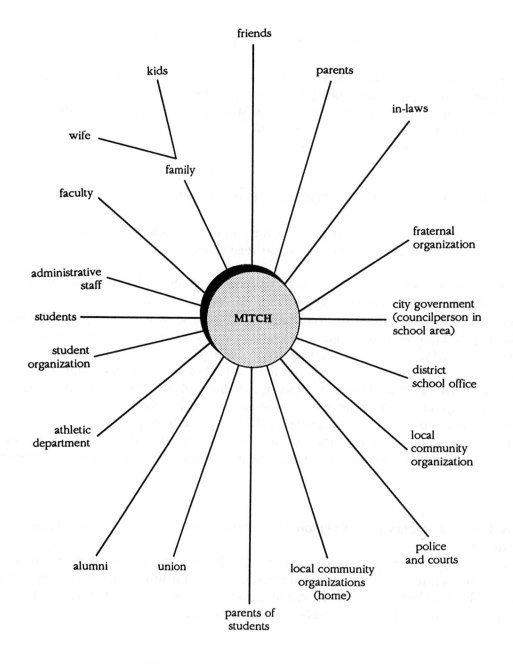

Figure 1. Chart of Social Settings

Family
- My wife says that I'm letting school consume me; she complains constantly because I don't spend enough time at home.
- My kids seem to withdraw from me because I'm a double authority figure, a father and a principal.

Parents
- My mother is infirm; my retired father calls me and tells me what a hard time he's having getting used to retirement.
- My mother tells me not to be spending time with her when I have so much to do, and then she complains to my wife and my father when I don't show up.

Mitch goes on to list demands, expectations, concerns, and frustrations that relate to each of the settings he has listed on his chart.

On a separate sheet of paper, list the demands, expectations, and concerns related to each of the social settings you have on your chart. Do not try to solve any of the problems you see cropping up. If some solution to a problem does suggest itself while you are doing this exercise, jot it down and put it aside.

3. **Identify conflicts between systems.** Once you review the expectations, demands, and concerns associated with each setting, list the conflicts between settings that cause you concern. Here are some of the conflicts Mitch identifies:

- My wife wants me to spend more time at home, and yet she criticizes me for not spending more time with my parents.
- The students, both individually and through their organizations, keep asking me to be more liberal while their parents are asking me to tighten things up.
- My administrative staff thinks that I'm taking sides against them in a dispute with the athletic department.
- My friends say that I spend so much time at work involving myself in crisis management that I have no time left for them; they tell me I'm doing myself in.

Conflicts Between the Social Settings of My Life

_____

_____

_____

_____

_____

_____

4. **Themes.** Share with the other members of your training group the kinds of conflicts you have identified. What kinds of themes emerge?

5. **Action.** Choose one conflict and identify some of the things you might do to resolve it. For instance, Mitch, in talking about the conflict between his work and his friends, decides to use the "overfull calendar" technique. That is, he puts social gatherings into his calendar. Then, when others ask for his time, he tells them what times are already taken and asks them to choose from the times that are left. Of course, he does not say that a certain day is taken up with an outing with friends.

a. **The conflict.** _____

_____

b. **Possible actions.** _____

_____

_____

_____

_____

LIFE SKILLS

Sometimes people develop problems or fail to manage them very well because they do not have the kinds of **life skills** needed to handle developmental tasks and to invest themselves effectively in the social systems of life. For instance, a young married couple, Roger and Tess, find that they don't have the communication skills needed to talk to each other reasonably about the problems they have been encountering during the first couple of years of marriage.

EXERCISE D: ASSESSING WORKING KNOWLEDGE AND SKILLS FOR EFFECTIVE LIVING

This exercise is a checklist designed to help you get in touch with both your resources and possible areas of deficit. Listed below are various groups of skills needed to undertake the tasks of everyday living. Rate yourself on each skill. The rating system is as follows:

5 I have a **very high** level of this skill.
4 I have a **moderately high** level of this skill.
3 From what I can judge, I am about **average** in this skill.
2 I have a **moderate deficit** in this skill.
1 I have a **serious deficit** in this skill.

You are also asked to rate how important each skill is in your eyes. Use the following scale.

5 For me this skill is **very important**.
4 For me this skill is of **moderate** importance.
3 For me this skill has **average** importance.
2 For me this skill is **rather unimportant**.
1 For me this skill is **not** important at all.

1. **Body-Related Skills**	Level	Importance
• Knowing how to put together nutritious meals.	____	____
• Knowing how to control weight.	____	____
• Knowing how to keep fit through exercise.	____	____
• Knowing how to maintain basic body hygiene.	____	____
• Basic grooming skills.	____	____
• Knowing what to do when everyday health problems such as colds and minor accidents occur.	____	____
• Skills related to sexual expression.	____	____
• Athletic skills.	____	____
• Aesthetic skills such as dancing.	____	____

Other body-related skills:

	Level	Importance
• _____	____	____
• _____	____	____
• _____	____	____

2. **Learning-How-to-Learn Skills**	Level	Importance
• Knowing how to read well.	____	____
• Knowing how to write clearly.	____	____
• Knowing basic mathematics.	____	____
• Knowing how to learn and study efficiently.	____	____
• Knowing something about the use of computers.	____	____
• Using history to understand today's events.	____	____
• Being able to use basic statistics.	____	____
• Knowing how to use a library.	____	____
• Knowing how to find information I need.	____	____

Other learning and learning-how-to-learn skills.

	Level	Importance
• _____	____	____
• _____	____	____
• _____	____	____

3. Skills Related to Values

	Level	Importance
• Knowing how to clarify my own values.	___	___
• Knowing how to identify the values of others who have a significant relationship to me.	___	___
• Knowing how to identify the values being "pushed" by the social systems to which I belong.	___	___
• Knowing how to construct and reconstruct my own set of values.	___	___

Other value-related skills:

	Level	Importance
• _____	___	___
• _____	___	___
• _____	___	___

4. Self-Management Skills

	Level	Importance
• Knowing how to plan and set realistic goals.	___	___
• Problem-solving or problem-management skills.	___	___
• Decision-making skills.	___	___
• Knowing and being able to use basic principles of behavior such as the use of incentives.	___	___
• Knowing how to manage my emotions.	___	___
• Knowing how to delay gratification.	___	___
• Assertiveness: knowing how to get my needs met while respecting the legitimate needs of others.	___	___

Other self-management skills:

	Level	Importance
• _____	___	___
• _____	___	___
• _____	___	___

5. Communication Skills

	Level	Importance
• The ability to speak before a group.	___	___
• The ability to listen to others actively.	___	___
• The ability to understand others.	___	___
• The ability to communicate understanding to others (empathy).	___	___
• The ability to challenge others reasonably.	___	___
• The ability to provide useful information to others.	___	___

	Level	Importance

- The ability to explore with another person what is happening in my relationship to him or her.   ___   ___

Other communication skills:

- _____   ___   ___

- _____   ___   ___

- _____   ___   ___

6. Skills Related to Small Groups

- Knowing how to be an effective, active member of a small group.   ___   ___
- Knowing how to design and organize a group.   ___   ___
- Knowing how to lead a small group.   ___   ___
- Team-building skills.   ___   ___

Other small-group skills:

- _____   ___   ___

- _____   ___   ___

- _____   ___   ___

Now that you have done a brief assessment of some life skills, indicate which skills, if improved, would help you manage your concerns, problems, or soft spots better. When you name a skill, indicate why such a skill is important to you and what you might do to develop it.

_____

_____

_____

_____

_____

_____

_____

What kinds of life skills do you think you need to become better at, not just to handle your own problems more effectively, but to be an effective counselor?

_____

_____

_____

_____

_____

_____

_____

_____

GENERAL STRENGTHS AND SOFT SPOTS

The next two sentence-completion exercises are designed to help you further review both soft spots and strengths.

EXERCISE E: A SENTENCE-COMPLETION ASSESSMENT OF PERSONAL PROBLEMS

Do these sentence-completion exercises quickly. They may help you expand in more specific ways what you have learned about yourself in the preceding exercises.

1. My biggest problem is _____

_____

2. I'm quite concerned about _____

_____

3. One of my other problems is _____

_____

4. Something I do that gives me trouble is _____

5. Something I fail to do that gets me into trouble is _____

_____

6. The social setting of life I find most troublesome is _____

_____

7. The most frequent negative feelings in my life are _____

_____

8. These negative feelings take place when _____

_____

9. The person I have most trouble with is _____

_____

10. What I find most troublesome in this relationship is _____

_____

11. Life would be better if _____

_____

12. I tend to do myself in when I _____

_____

13. I don't cope very well with _____

_____

14. What sets me most on edge is _____

_____

15. I get anxious when _____

_____

16. A value I fail to put into practice is _____

17. I'm afraid to _____

_____

18. I wish I _____

_____

19. I wish I didn't _____

_____

20. What others dislike most about me is _____

_____

21. What I don't seem to handle well is _____

_____

22. I don't seem to have the skills I need in order to _____

_____

23. A problem that keeps coming back is _____

_____

24. If I could change just one thing in myself it would be _____

_____

EXERCISE F: A SENTENCE-COMPLETION ASSESSMENT OF STRENGTHS

1. One thing I like about myself is _____

_____

2. One thing others like about me is _____

_____

3. One thing I do very well is _____

4. A recent problem I've handled very well is _____

_____

5. When I'm at my best I _____

_____

6. I'm glad that I am able to _____

_____

7. Those who know me know that I can _____

_____

8. A compliment that has been paid to me recently is _____

_____

9. A value that I try hard to practice is _____

_____

10. An example of my caring about others is _____

_____

11. People can count on me to _____

_____

12. They said I did a good job when I _____

_____

13. Something I'm handling better this year than last is _____

_____

14. One thing that I've overcome is _____

_____

15. A good example of my ability to manage my life is _____

151

16. I'm best with people when _____

_____

17. One goal I'm presently working toward is _____

_____

18. A recent temptation that I managed to overcome was _____

_____

19. I pleasantly surprised myself when I _____

_____

20. I think that I have the guts to _____

_____

21. If I had to say one good thing about myself I'd say that I _____

_____

22. One way I successfully control my emotions is _____

_____

23. One way in which I am very dependable is _____

_____

24. One important thing I intend to do within the next two months is _____

_____

EXERCISE G: USING STRENGTHS TO MANAGE SOFT SPOTS

Sometimes we fail to see that our strengths provide the resources we need to manage areas in our lives needing improvement. In this exercise you are asked to relate resources to problems or undeveloped opportunities in your life.

1. List two of your key concerns or problems.

2. After each, list the resources you've identified in yourself that can help you manage your problem more effectively. Try to name resources that you are not currently using to manage the particular problem area.

Example: Katrina, a trainee in her late 20s, is concerned about bouts of anxiety. She has led a rather sheltered life and now realizes that she needs to "break out" in a number of ways if she is to be an effective counselor. The fact that she has been sheltered from so much is at the root of her anxiety. The counseling program has put her in touch with all sorts of people, and this is anxiety provoking. She lists some of the strengths she has that can help her overcome her anxiety.

- I'm bright. I know that a great deal of my anxiety comes from fear of the unknown.
- I'm intellectually adventuresome. I pursue new ideas. Perhaps this can help me be more adventuresome in seeking new experiences. New ideas don't kill me. I bet new experiences won't either.
- People find me easy to talk to. This can help me form friendships. My fears have kept me from developing friendships. However, developing more relationships is the royal route to the kinds of experiences I need.
- At root, I'm a religious person. This can help me put my fears into a larger context. In the larger context they seem petty.

Apply the same procedure to your two concerns.

Concern #1. _____

_____

Strengths that can be used to manage this concern. _____

_____

_____

Concern #2. _____

_____

Strengths that can be used to manage this concern. _____

_____

_____

APPENDIX TWO

SUPPLEMENTAL EXERCISES

These supplemental exercises are offered to trainees who believe that they would profit from doing even more items in various exercises. The numbers of the exercises correspond to the numbers in the main part of the manual.

EXERCISE 9: LISTENING TO KEY EXPERIENCES, BEHAVIORS, AND FEELINGS

A. A woman, 53, about to get divorced: "My husband and I just decided to get a divorce. (Her voice is very soft, her speech is slow, halting.) I really don't look forward to the legal part of it (pause) to *any* part of it to tell the truth. For the first time in my life I just sit around and think a lot. Or stare into space. I just don't know what to expect. (She sighs heavily.) I'm well into middle age. I don't think another marriage is possible. I just don't know what to expect."

a. **Client's key experiences:**

_____

_____

b. **Client's key behaviors:**

_____

_____

c. **What feelings/emotions do these experiences and behaviors generate?**

_____

_____

B. A man, 45, with a daughter, 14, who was just hit by a car: "I should never have allowed my daughter to go to the movies alone. (He keeps wringing his hands.) I don't know what my wife will say when she gets home from work. (He grimaces.) She says I'm careless . . . but being careless with the kids . . . that's something else! (He stands up and walks around.) I almost feel as if *I* had broken Karen's arm, not the guy in that car. (He sits down, stares at the floor, keeps tapping his fingers on the desk.) I don't know."

a. **Client's key experiences:**

_____

_____

b. **Client's key behaviors:**

_____

_____

c. **What feelings/emotions do these experiences and behaviors generate?**

_____

_____

C. A senior in high school, 17, talking to his girl friend: "My teacher told me today that I've done better work than she ever expected. I always thought I could be good at my studies if I applied myself. (He smiles.) So I tried this semester, and it's paid off. It's really paid off!"

a. **Client's key experiences:**

_____

_____

b. **Client's key behaviors:**

_____

_____

c. **What feelings/emotions do these experiences and behaviors generate?**

_____

_____

D. A trainee, 29, speaking to the members of his training group: "I don't know what to expect in this group. (He speaks hesitatingly.) I've never been in this kind of group before. From what I've seen so far, I, well, I get the feeling that you're pros, and I keep watching myself to see if I'm doing things right. (Sighs heavily.) I'm comparing myself to what everyone else is doing. I want to get good at this stuff . . . (pause) . . . but frankly I'm not sure I can make it."

a. **Client's key experiences:**

_____

_____

b. **Client's key behaviors:**

_____

_____

c. **What feelings/emotions do these experiences and behaviors generate?**

_____

_____

E. A young women, 20, speaking to a college counselor toward the end of her second year: "I've been in college almost two years now, and nothing much has happened. (She speaks listlessly.) The teachers here are only so-so. I thought they'd be a lot better. At least that's what I heard. And I can't say much for the social life here. Things go on the same from day to day, from week to week."

a. **Client's key experiences:**

_____

_____

b. **Client's key behaviors:**

_____

_____

c. **What feelings/emotions do these experiences and behaviors generate?**

_____

_____

F. A woman, 42, married, with three children in their early teens, speaking to a church counselor: "Why does my husband keep blaming me for his trouble with the kids? I'm always in the middle. He complains to me about them. They complain to me about him. (She looks the counselor straight in the eye and talks very deliberately.) I could walk out on the whole thing right now. Who the hell do they think they are?"

a. **Client's key experiences:**

_____

_____

b. **Client's key behaviors:**

_____

_____

c. **What feelings/emotions do these experiences and behaviors generate?**

_____

_____

G. A bachelor, 39, speaking to the members of a life-style group to which he has belonged for about a year: "I've finally met a woman who is very genuine and who lets me be myself. I can care deeply about her without making a child out of her. (He is speaking in a soft, steady voice.) And she cares about me without mothering me. I never thought it would happen. (He raises his voice a bit.) Is it actually happening to me? Is it actually happening?"

a. **Client's key experiences:**

_____

_____

b. **Client's key behaviors:**

_____

_____

c. **What feelings/emotions do these experiences and behaviors generate?**

_____

_____

EXERCISE 12: COMMUNICATING UNDERSTANDING OF A CLIENT'S FEELINGS

A. This young man has just been abandoned by his wife: "We've been married for about a year. She left a note saying that 'this has not been working out.' Just like that. I thought that things were going fairly well. I don't know what I can do to get her back. Can you make someone love you?"

_____

_____

_____

_____

B. This woman has been suffering from migraine headaches for a long time: "They seem to be getting worse. So far nothing has helped me reduce their number or to manage them once they start. I won't go to a doctor. They've never helped. Oh, I read about those so-called 'new' treatments, but I bet it's the same old stuff promising you a lot and delivering nothing."

_____

_____

_____

_____

C. This man is talking about having to work two jobs to support his family: "I guess I'm fortunate to have both jobs, but I've got no time for myself. The jobs eat into my evening hours and the weekends. I think what really bothers me is that my family does not seem to notice my absence. They take for granted all the hours I've been putting in. It's as if life is about nothing else but work, and no one else cares."

_____

_____

_____

_____

EXERCISE 13: USING A FORMULA TO EXPRESS EMPATHY

A. Man, 35, talking to a counselor at the company where he works (he wrings his hands as he talks): "I'm going to the hospital tomorrow for some tests. The doctor suspects an ulcer. But nobody has told me exactly what kind of tests. I'm supposed to take these enemas and not eat anything tomorrow. I've heard rumors about what these tests are like, but I don't really know."

B. Woman, 28, talking about her job with a colleague: "It's not a big thing. But this is the third time this month I've been asked to change hours with her. It certainly seems to indicate who is more important there. Why does it always have to be me who defers to her?"

C. Student, 16, talking to a teacher whom he trusts about a teacher with whom he has not been getting on: "I thought he was going to really chew me out. I was afraid that he was just going to tell me he was kicking me out of class. But we sat in his office and talked about our differences!"

D. Graduate student, 25, to adviser: "I have two term papers due tomorrow. I'm giving a report in class this afternoon. My wife is down with the flu. And now I find out that a special committee wants to 'talk' with me about my 'progress' in the program."

E. Trainee, 21, in a counselor education program to a fellow trainee: "I know he's going to ask me to counsel someone before the whole group. I can't even imagine myself standing up there! I actually feel like skipping the class and telling him I was sick."

EXERCISE 14: USING YOUR OWN WORDS TO EXPRESS EMPATHY

A. A high school counselor, 41, talking to a colleague: "Sometimes I think I'm living a lie. I don't have any interest in high school kids anymore. So when they come into my office, I don't really do much to help them. Most of them and their problems bore me. But I've been here now for twelve years. I like living around here. I try half-heartedly to work up some interest, but I don't get far."

a. Use the formula.

_____

_____

b. Use your own words.

_____

_____

_____

B. A man, 35, who has not been feeling well, talking to a friend who is a nurse: "I'm going into the hospital tomorrow for some tests. I think they suspect an ulcer. (He fidgets.) But nobody has told me exactly what kind of tests. I'm supposed to take these enemas and not eat anything after supper this evening. I've heard rumors about these kinds of tests, but I'm not really sure what they're like. I'm not even sure that I want to know."

a. Use the formula.

_____

_____

b. Use your own words.

_____

_____

_____

C. A graduate student, 25, to her adviser: "I have two term papers due tomorrow. I'm giving a report in class this afternoon. My husband is down with the flu. And now I find out that the student committee wants to 'talk' with me about my 'progress' in the program. I think the last straw must be around here someplace."

a. Use the formula.

_____

_____

b. Use your own words.

_____

D. A woman, 43, talking to a counselor in a rape crisis center: "It was all I could do to come here. A friend told me to call the police. I didn't. I didn't want to become one of those stories you read in the paper every day! They'd be asking me all sorts of questions. And hinting that it was probably my fault. Ugh! I just want to forget it. I don't want to keep reliving it over and over again."

a. Use the formula.

_____

_____

b. Use your own words.

_____

_____

_____

E. A female high school student, 17, talking to a male counselor about an unexpected pregnancy: "I, well, I don't think I can talk about it here. . . (pause). . . You being a man and all that. (pause) What happens between me and my boyfriend and me and my family—well, that's all very personal. I don't talk to strangers about personal things."

a. Use the formula.

_____

_____

b. Use your own words.

_____

_____

_____

Since empathy is such a key communication skill for helpers, it will be revisited in subsequent pages in conjunction with other skills such as probing and challenging.

EXERCISE 22: COMBINING EMPATHY WITH PROBES FOR CLARITY AND CLIENT ACTION

A. A divorced woman, 44, talking to a counselor about her drinking. This is the second session. She has spent a lot of time telling her story. To the counselor there seemed to be a lot of evasions and some outright lying: "Actually, it's a relief to tell someone. I don't have to give you any excuses or make the story sound right. I drink because I like to drink; I'm just crazy about the stuff, that's all. But I'm under no delusions that telling you is going to solve anything. When I get out of here, I know I'm going straight to a bar and drink. Some new bar, new faces, some place they don't know me."

a. Empathy.

_____

_____

b. Fruitful area for probing.

_____

_____

c. Probe.

_____

_____

d. Probes for action possibilities.

_____

_____

_____

B. A man, 57, talking to a counselor about a family problem: "My younger brother, he's 53, has always been a kind of bum. He's always poaching off the rest of the family. Last week my unmarried sister told me that she'd given him some money for a 'business deal.' Business deal, my foot! I'd like to get hold of him and kick his ass! Oh, he's not a vicious guy. Just weak. He's never been able to get a fix on life. But he's got the whole family in turmoil now, and we can't keep going through hell for him."

a. Empathy.

_____

b. Fruitful area for probing.

_____

_____

c. Probe.

_____

_____

d. Probes for action possibilities.

_____

_____

_____

C. A woman, 49, talking to a counselor about her relationship with her husband: "To put it frankly, my husband isn't very interested in me sexually anymore. We've had sex maybe once or twice in the last two or three months. What makes it worse is that I still have very strong sexual feelings. It seems they're even stronger than they used to be. I keep thinking about this all the time. He doesn't seem very interested at all. I don't know if he's got someone on the side. I'm not handling it well."

a. Empathy.

_____

_____

b. Fruitful area for probing.

_____

_____

c. Probe.

_____

_____

163

d. Probes for action possibilities.

_____

_____

_____

D. A man, 49, talking to a rehabilitation counselor after an operation that has left him with one lung: "I'll never be as active as I used to be. But at least I'm beginning to see that life is still worth living. I have to take a long look at the possibilities, no matter how much they've narrowed. I can't explain it, but there's something good stirring in me."

a. Empathy.

_____

_____

b. Fruitful area for probing.

_____

_____

c. Probe.

_____

_____

d. Probes for action possibilities.

_____

_____

_____

E. Mark and Lisa, a married couple, both 33, after years of attempting to have children finally adopted a baby girl, Andrea. Their relationship, which up to then seemed quite good, has begun to disintegrate. Mark has made some cracks about "the stronger one in the house." Andrea has proved to be a somewhat difficult baby. Lisa feels exhausted and blames Mark for not helping her. They are both thinking about divorce now but feel very guilty because of the child. Both of them say, "If only we had never adopted Andrea."

a. Empathy.

_____

_____

b. Fruitful area for probing.

_____

_____

c. Probe.

_____

_____

d. Probes for action possibilities.

_____

_____

_____

EXERCISE 26: INFORMATION LEADING TO NEW PERSPECTIVES AND ACTION

A. Cindy, 23, has been bleeding internally. She has been told that she needs to undergo a series of tests. She is very frightened and fears the worst. She has never been seriously sick in her life. She fears the doctors, the tests, the hospital. She has never even visited anyone in a hospital. She does not want to go.

What information might help this client develop new perspectives? What actions might these new perspectives lead to?

_____

_____

_____

_____

B. Tim, 18, has been smoking marijuana for about three years. He is a fairly heavy user. He has recently received several shocks. His father died suddenly and his steady girlfriend left him.

165

Since he was not especially close to his father, he is surprised by how hard he is hit by the loss. After reading a couple of articles on marijuana use, he developed fears that he has been doing irreversible genetic damage to himself. His is fearful of giving it up because he thinks he needs it to carry him over this period of special stress and because he fears withdrawal symptoms.

What information might help this client develop new perspectives? What actions might these new perspectives lead to?

_____

_____

_____

_____

C. Edna, 17, is an unmarried woman who is experiencing her third unexpected and unwanted pregnancy. The other two ended in abortion. She feels guilty about the third unwanted pregnancy and about the abortions. Many of her relatives belong to a fundamentalist church. She is thinking about keeping this child. For all her promiscuity, she seems to know little about sex. She believes that men just take advantage of her.

What information might help this client develop new perspectives? What actions might these new perspectives lead to?

_____

_____

_____

_____

D. Maxine, 54, has suffered a stroke that has left her partially paralyzed on her left side and with speech that is a bit slurred. Her husband, who seems lost without her, seems artificially cheerful during his visits. She is about to be transferred to a rehabilitation unit. She has a long road ahead of her. She is depressed.

What information might help this client develop new perspectives? What actions might these new perspectives lead to?

_____

_____

_____

EXERCISE 28: THE DISTINCTION BETWEEN BASIC AND ADVANCED EMPATHY

A. A high school senior is talking to a school counselor about college and what kinds of courses she might take there. However, she also mentions, somewhat tentatively, her disappointment in not being chosen as valedictorian of her class. She and almost everyone else had expected her to be chosen. She says: "I know that I would have liked to have been the class valedictorian, but I'm not so sure that you are supposed to count on anything like that. They chose Jane. She'll be good. She speaks well, and she's very popular. But no one has a right to be valedictorian. I'd be kidding myself if I thought differently. I've done better in school than Jane, but I'm not as outgoing or popular."

a. Basic empathy.

_____

_____

b. Your hunch and your reason for it.

_____

_____

_____

c. Advanced empathy.

_____

_____

_____

B. A college professor, 43, is talking to a friend, who happens to be a counselor, about his values. He is vaguely dissatisfied with his priorities but has never done much about examining his current values in any serious way. From time to time the two of them talk about values, but no conclusions are reached. He is not married. Work seems to be a primary value. He says: "Well, it's no news to you that I work a lot. There's literally no day I get up and say to myself, 'Well, today is a day off and I can just do what I want.' It sounds terrible when I put it that way. I've been going on like that for about ten years now. It seems that I should do something about it. But it's obviously my choice. I'm doing what I doing freely. No one's got a gun to my head."

a. Basic empathy.

_____

b. Your hunch and your reason for it.

_____

_____

_____

c. Advanced empathy.

_____

_____

_____

_____

C. A man, 50, with a variety of problems in living is talking with a counselor. His tendency has been to ruminate almost constantly on his defects. He begins a second interview on this somewhat sour note. He says: "To make myself feel bad, all I have to do is review what has happened to me in the past and take a good look at what is happening to me right now. This past year, I let my drinking problem get the best of me for four months. Over the years, I've done lots of things to mess up my marriage. For instance, like changing jobs all the time. Now my wife and I are separated. I don't earn enough money to give her much, and the thought of getting another job is silly with the economy the way it is. I'm not so sure what skills I have to market, anyway."

a. Basic empathy.

_____

_____

b. Your hunch and your reason for it.

_____

_____

_____

c. Advanced empathy.

_____

_____

_____

_____

D. A divorced woman, 35, with a daughter, 12, is talking to a counselor about her current relationship with men. She mentions that she has lied to her daughter about her sex life. She has told her that she doesn't have sexual relations with men, but she does. In general she seems quite protective of her daughter. From what her mother says, however, the young girl does not seem to have any serious problems. She says: "I don't want to hurt my daughter by letting her see my shadow side. I don't know whether she could handle it. What do you think? I'd like to be honest and tell her everything. I just don't want her to think less of me. I like sex. I've been used to it in marriage, and it's just too hard to give it up. I wish you could tell me what to do about my daughter."

a. Basic empathy.

_____

_____

b. Your hunch and your reason for it.

_____

_____

_____

c. Advanced empathy.

_____

_____

_____

E. The wife of this man, 35, has recently left him. He tried desperately to get her back, but she wanted a divorce. As part of his strategy to get her back, he examined his role in the marriage and freely "confessed" to both the counselor and his wife what he felt he was doing wrong in the relationship. Part of his problem in his relationships with both his wife and others was a need to get the better of her and others in arguments. He could never admit that he might have been wrong. He says: "I don't know what's wrong with her. I've given her everything she wanted. I mean I've admitted all my mistakes. I was even willing to take the blame for things that I

169

thought were her fault. But she's not interested in a reformed me! Damned if you do, damned if you don't."

a. Basic empathy.

_____

_____

b. Your hunch and your reason for it.

_____

_____

_____

c. Advanced empathy.

_____

_____

_____

_____

EXERCISE 33: RESPONDING TO SITUATIONS CALLING FOR IMMEDIACY

A. **The situation.** The client is a male, 22, who is obliged to see you as part of being put on probation for a crime he committed. He is cooperative for a session or two and then becomes quite resistant. His resistance takes the form of both subtle and not too subtle questioning of your competence, questioning the value of this kind of helping, coming late for sessions, and generally treating you like an unnecessary burden.

Immediacy response.

_____

_____

_____

_____

_____

_____

_____

B. **The situation.** You are a woman. The client, 19, reminds you of your own son, 17, toward whom you have mixed feelings as he struggles to establish some kind of reasonable independence from you. The client at times acts in very dependent ways toward you, telling you that he is glad that you are helping him, asking your advice, and in various ways taking a "little boy" posture toward you. At other times he seems to wish that he didn't have anything to do with you at all and accuses you of being "like his mother."

Immediacy response.

_____

_____

_____

_____

_____

_____

_____

_____

EXERCISE 43: MAKING GOALS SPECIFIC

A. Nancy, 19, unmarried, is facing the problem of an unwanted pregnancy. She has a variety of problems. Her parents are extremely upset with her. Her father won't even talk to her. She lives at home and is attending a local community college. These living arrangements are now unsatisfactory to her. Since, for value reasons, she has decided against an abortion, she does not want to live out the remaining months of pregnancy in an atmosphere of hostility and conflict. She is upset because her education is going to be interrupted, and finishing college has always been high on her list of priorities. She is unsure about her finances and resents being financially dependent on her parents.

Statement of good intention.

_____

_____

Broad goal.

_____

_____

_____

Specific goal.

_____

_____

How is it to be measured? How will we know that it has been accomplished?

_____

_____

B. Julian, 51, a man separated from his wife for seven years, has just lost a son, 19, in an automobile accident. He (Julian) was driving with his son when they were struck by a car that veered into them from the other side of the road. Julian, who had his seat belt fastened, escaped with only cuts and bruises. His son was thrown through the windshield and killed instantly. The driver of the other car is still in critical condition and may or may not live. Now, ten days after the accident, Julian is still in psychological shock and plagued with anger, guilt, and grief. He has not gone back to work and has been avoiding relatives and friends because he finds getting sympathy "painful."

Statement of good intention.

_____

_____

Broad goal.

_____

_____

_____

Specific goal.

_____

How is it to be measured? How will we know that it has been accomplished?

_____

_____

C. Felicia, 44, finds that her nonassertiveness is causing her problems. She is especially bothered at work. She finds that a number of people in the office feel quite free to interrupt her when she is in the middle of a project. She gets angry with herself because her tendency is to put aside what she is doing and try to meet the needs of the person who has interrupted her. As a result, she sometimes misses important deadlines associated with the projects on which she is working. She feels that others see her as a "soft touch."

Statement of good intention.

_____

_____

Broad goal.

_____

_____

_____

Specific goal.

_____

_____

How is it to be measured? How will we know that it has been accomplished?

_____

_____

To the Owner of This Book

I hope that you have enjoyed *Exercises in Helping Skills* (Fifth Edition). I'd like to know as much about your experiences with these exercises as possible. Only through your comments and the comments of others can I learn how to make this a better book for future readers.

School: _____ Your Instructor's Name: _____

1. What I like *most* about these exercises is: _____

_____

2. What I like *least* about these exercises is: _____

_____

3. My specific suggestions for improving these exercises are: _____

_____

4. Some ways in which I used these exercises in class were: _____

_____

5. Some ways in which I used these exercises out of class were: _____

_____

6. Some of the exercises that were used most meaningfully in my class were: _____

_____

7. My general reaction to these exercises is: _____

8. In the space below or in a separate letter, please write any other comments about the book you'd like to make. I welcome your suggestions!

_____

_____

_____

9. Please write down the name of the course in which you used *Exercises in Helping Skills*.

_____

Brooks/Cole is dedicated to publishing quality publications for education in the human services fields. If you are interested in learning more about our publications, please fill in your name and address and request our latest catalogue.

Name _____

Street Address _____

City, State, and Zip _____

FOLD HERE

BUSINESS REPLY MAIL

FIRST CLASS   PERMIT NO. 358   PACIFIC GROVE, CA

POSTAGE WILL BE PAID BY ADDRESSEE

ATT: Human Services Catalogue

Brooks/Cole Publishing Company
511 Forest Lodge Road
Pacific Grove, California 93950-9968

FOLD HERE